THE TRANSITION TO FLEXIBILITY

THE TRANSITION TO FLEXIBILITY

Edited by
Daniel C. Knudsen

Department of Geography &
Institute for Development Strategies
School of Public and Environmental Affairs
Indiana University at Bloomington
Bloomington, Indiana 47401

Kluwer Academic Publishers
Boston/London/Dordrecht

Distributors for North America:
Kluwer Academic Publishers
101 Philip Drive
Assinippi Park
Norwell, Massachusetts 02061 USA

Distributors for all other countries:
Kluwer Academic Publishers Group
Distribution Centre
Post Office Box 322
3300 AH Dordrecht, THE NETHERLANDS

Library of Congress Cataloging-in-Publication Data

The transition to flexibility / edited by Daniel C. Knudsen.
p. cm.
Includes bibliographical references.
ISBN 0-7923-9760-6
1. Industrial organization. 2. Production management. 3. Market segmentation. 4. Technological innovations. 5. Organizational change. I. Knudsen, Daniel C., 1955-
HD31.T676 1996
658.5--dc20 96-9476
CIP

Printed on acid-free paper.

Printed in the United States of America

CONTENTS

3 Labor and Capitalist Accumulation in the Late Twentieth Century: What and Who is Flexible?
Martin Kenney and *James Curry*

4 Preparing Workers and Students for the New Workplace
Larry Mikulecky

5 Large Firm Strategies: Spatial Patterns of Production
Claire Ellen Pavlik

6 Services and Flexible Production: Contingent But Mutually Reinforcing Systems

Barney Warf

7 Inter-Firm Collaboration and Industrial Competitiveness

Brian Bosworth

8 Government Response to Structural Change and the Implementation of More Flexible Management and Production Systems

J.W. Wheeler

LISTS OF TABLES AND FIGURES

Tables

Figures

ACKNOWLEDGEMENTS

The essays in this volume originated as series of papers presented at a conference on flexible accumulation held in Bloomington, Indiana in June of 1994. The conference would not have been possible without the support of Ameritech Professor for Economic Development Charles F. Bonser. I would also like to thank Ray Martin, Betty Fiscus, and Nancy Croker for valuable support. Special thanks are due Cynthia Mahigian Moorhead for making the manuscript camera-ready. Finally, my thanks to two anonymous referees and to the staff at Kluwer Academic for suggestions and guidance in bringing this project to completion.

Naturally any remaining mistakes and oversights are the responsibility of the authors.

1

THE CURRENT TRANSITION IN INDUSTRIAL CAPITALISM

Daniel C. Knudsen
Department of Geography and Institute for Development Strategies, School of Public and Environmental Affairs, Indiana University, Bloomington, IN 47405 USA
and
Jeffrey S. Boggs
Department of Geography, Indiana University, Bloomington, IN 47405 USA

INTRODUCTION

The advanced capitalist nations are currently undergoing a wrenching economic, social, and political transformation. At its heart, this transformation involves a shift away from the mass production of highly standardized goods and services for mass markets to small-batch production of relatively customized goods and services for niche markets. This volume brings together in one place essays by a group of individuals who have thought widely and written extensively on various aspects of the current transformation of industrial capitalism in an attempt to integrate this research and these disparate viewpoints into an single complete picture. Briefly, the book touches upon five dimensions of the current transformation of capitalism: technology, labor relations, firm organization, public policy, and locational decisionmaking. Each chapter is concerned with a different aspect of the same transformative process. The purpose of this introductory chapter is to frame the debate contained within the remainder of the book. To do so, it is first necessary to define a set of terms.

Throughout this book, the terms "Fordist manufacturing" and "Fordist production" appear. These terms refer to the particular form of industrial production characterized by the mass manufacturing of relatively standardized products by large, vertically integrated firms. The terms "Fordism" and "Fordist accumulation" refer to the accompanying regime of accumulation, that is, the mode of production along with its accompanying social and political institutions. Spatially, Fordism is typically identified with national systems of capitalism.

"Flexible manufacturing" and "flexible production" refer to the particular form of industrial production characterized by small batch manufacturing of relatively customized products by highly focussed, vertically disintegrated firms. The terms "flexibilism" and "flexible accumulation" refer to flexible production along with its emerging social and political institutions. Spatially, flexible accumulation is typically identified with international or global capitalism.

TRANSITIONS IN INDUSTRIAL CAPITALISM: A REGULATIONIST APPROACH

Terms such as "Fordism" and "flexibilism" have their origins in regulation theory, a framework of political economy derived from historical materialism, or "Marxism" (seminal regulation theory works include Aglietta, 1979; Lipietz, 1986; Boyer, 1986, 1988, 1990; Dunford, 1990; Jessop, 1990). "'Regulation' refers to maintaining the institutional fabric of [capitalist] growth in a dynamic and contradictory setting through state interventions and class compromises" (Storper and Walker, 1989, p. 203). Regulation theory provides a useful framework for examining the technology, labor relations, firm organization, public policy, and geography associated with historical periods of industrial capitalism. Regulation theory provides a number of advantages over previous political-economic frameworks. Unlike modernization and dependency theories, regulation theory is not based on limited socio-cultural experience; unlike Althusserian structuralism, social and political institutions play an active and central role in the regulationist framework and are not strictly derivative of economic logic alone. To existing Marxian concepts of mode of production and crisis, regulationists add concepts of mode of (social) regulation, and regime of accumulation, and they substantially redefine Marxian notions of geography and crisis.

Mode of Production

The Marxian definition of the mode of production includes both the organization of workers and applied technologies. It is an abstract system of social relations organizing the forces of production in the transformation of nature (Peet, 1991, p. 62). Peet notes that any capitalist mode of production has two basic social objectives. The first being the production of material goods used to reproduce labor power. The second is the creation of a surplus of products, intended for one of two uses: investment in new means of production or to reproduce the capitalist class (Peet, 1991, p. 62).

Within the mode of production, regulation theory identifies a dualistic relationship involving technological ensembles and engendered capital/labor relations as the driving force behind successive regimes of capitalist accumulation (for example Scott, 1988). Regulation theorists adhere to the notion of Marx (1977, pp. 429-438) that the capitalist dynamic is primarily fueled by technological innovation. However, they are quick to point out that it is not innovation *per se*, but technological ensembles (groups of new, profitable, and compatible technologies) that are the driving force behind the capitalist dynamic. According to regulation theory, new technologies, however, also revolutionize the capital/labor relationship in the workplace, since they alter the demands that capital places on labor. Further, Clark (1989, p. 9) argues that these new workplace relationships in turn alter social institutions within the community (and social relations within the community alter workplace relationships), with social institutions changing in accordance with the needs of capital.

Mode of Social Regulation

The mode of social regulation, sometimes referred to simply as a mode of regulation, is a system of social and political institutions which partially equilibrate production and consumption. The mode of social regulation coherently focuses the roles of political and social relations, which include state action and social institutions, behavioral norms and habits, and political practices, to insure capital accumulation.

Because capitalism has no internal and automatic mechanism insuring that "workers adjust their work habits, training, attitudes, location, and so on, to socio-technical change, or that consumers adjust their purchases to growing social surpluses, regimes of accumulation must be given coherence by social processes and institutions that collectively constitute a 'mode of regulation'" (Kenney and Florida, 1989, p. 138). Modes of regulation are generated piece-wise through political struggles and compromises. "Economic institutions evolve as a process of response and adjustment to political conflict" (Kenney and Florida, 1989, p. 138). Thus, the reproduction of capitalism is never guaranteed, but rather must be continually secured and resecured through a range of social norms, mechanisms, and institutions that help temporarily stabilize capitalism around a particular regime of accumulation.

Regime of Accumulation

Both the mode of production and the mode of social regulation combine to form a regime of accumulation, which is "the overall productive system" (Teague, 1990, p. 39). Regimes of accumulation are defined here as historical periods of capital accumulation characterized by relative social stability and economic growth. Accordingly, regulationist work focuses on variations of the capitalist mode of production, such as the putting-out system, manufactory, machinofacture, Fordism, and post-Fordism or flexibilism. At the same time, since a society's economic structure is defined by a complex system of articulated modes of production, distinctly different modes of production often exist within or adjacent to the same social spaces (Boyer, 1990, p. 33).

Geography

Geography poses a particular problem for regulation theory. On the one hand "each regime of accumulation exhibits a distinctive economic geography" (Tickell and Peck, 1992, p. 197). On the other hand, only through a series of refinements has the original theory been made to yield sufficient insights geographically. As originally proposed, regulation theory assumes that the mechanisms and components of regulation that operate at the national level are somehow translated locally without any real explanation of how that occurs (Goodwin, Duncan, and Halford, 1993). Numerous authors (notably Jameson, 1991; Sayer, 1989; Massey, 1984; Goodwin, Duncan, and Halford, 1993; Jessop, Nielsen, and Pedersen, 1993; Clark, 1989) have pointed out that social norms, mechanisms, and institutions do not exist in a vacuum, but are themselves outcomes of distinct local, regional, national, and global history. Local history mediates national and global capitalism, and local agencies translate national practices and relations of regulation via local regulatory institutions and mechanisms. Indeed, the exact division of labor between different tiers and agencies of the state (national, state, local) is open to constant contest and negotiation, and the resultant allocation of state functions is driven by a combination of political, social, cultural, and economic considerations (Goodwin, Duncan, and Halford, 1993, p. 69). The local state is thus at once an agent and an obstacle to the central (national) state, and its role is contextual.

The result of this global/local nexus is that, despite a generally synchronized international crisis encompassing essentially identical changes in production, institutions and class relations, economic differentiation occurs subnationally because of locally varying social forces and cultural practices, and because local agencies are the medium through which regulatory practices are both interpreted and delivered (Goodwin, Duncan, and Halford, 1993). Further, because technology can change much more rapidly and is much less localized than its associated institutional arrangements, new technology is likely to be applied in regions where it is least circumscribed by an existing set of institutions and least likely to be applied in regions where redundant technologies (and their associated institutions) are most entrenched. It is this concomitant positioning of technology and institutions that leads to the spatial unevenness of the development process.

This notion is perhaps most clearly articulated in the literature on new industrial spaces (Scott, 1988; and Storper and Walker, 1989 are considered seminal works), although it is not foreign to more recent writings on long waves of capitalist development (Mandel, 1980; Rothwell, 1982; Gruber and Nowotny 1990; Berry, 1991). Hall (1985, p. 10), for example, notes that "new industrial traditions . . . [take] root in places different from older ones" implying that as leading sectors of a regime decline in importance, so do the regions that contain them. Storper and Scott (1989) and Storper and Walker (1989) also take pains to point out that emergent industries do not arise phoenix-like from the ashes of old modes of production.

Misregulation and Crisis

Regulation theory focuses on the long-term possibilities of capital accumulation and recognizes the moderating influence of institutions in preventing a short-term economic collapse. However, these dynamic, short-run adjustments toward equilibrium induce a natural long-term disequilibrium into the economic system. These dynamics lead to a point where regulation is no longer possible in the current social system. This spawns a crisis, which leads to a radical restructuring of the production, social and spatial systems (Boyer, 1990, pp. 48-60).

Misregulation occurs when the mode of production and the mode of social regulation begin to impede the process of capital accumulation. It manifests itself as disequilibria between production and consumption. Changes in consumptive and productive norms—whether technologically or socially grounded—are necessary before some measure of stability returns. As such, misregulation often spurs incremental change in the mode of production, the mode of social regulation, or both as social institutions attempt to maintain capital accumulation. Misregulation can also spawn an endogenous crisis within a regime of accumulation, if it eventually blocks capital accumulation (Boyer, 1990, pp. 48-60).

Crises occur due to misregulation (endogenous) or external calamities (exogenous). Endogenous crises result from conflicts within the regime of accumulation which can no longer be mediated, ignored, or squelched by social institutions. Often an endogenous crisis results from a culmination of misregulations, with the last event acting as the proverbial "nail in the coffin." Innovation typically plays a central role in endogenous crisis generation, often undermining a regime of accumulation.

In the words of Kenney and Florida:

> According to regulation theory, economic crises are initially signaled when the organization of production runs up against its own internal limits. This inability to realize productivity increases sets in motion a growing divergence between the ability to generate output or social product and to allocate appropriate shares to accumulation and consumption. During such periods the arrangement of the economy hardens, blocking the kinds of institutional reorganization necessary for restructuring to occur. The solution to crisis is two-fold. On one hand, development of new technologies and/or reorganization of work is needed to create the conditions for a new round of economic expansion. On the other hand, only social conflict can bring about the kinds of institutional restructuring needed to set such growth in motion. In effect, social forces provide the motor force of economic change; they must be harnessed before another period of expansion can begin (1989, p. 138).

Exogenous crises result from natural disasters and war. Exogenous crises occur less predictably and more intensely than endogenous crises.

FORDISM

The capitalist growth dynamic, as recognized by regulation theory, incorporates two sets of structures: the mode of production and the mode of social regulation. The mode of production includes industrial capital (and indirectly technology), the capital/labor relation on the production floor, intrafirm management, and interfirm linkages and networks. The mode of social regulation as recognized by regulation theory incorporates the relationship between work and class structure, the interaction of work, class and gender, mechanisms for the reproduction of class, and political institutions (both government programs that aid in the reproduction of capital and labor, and the formal bureaucracy used to carry out those programs). All this is embedded within a spatio-historical context in which capital, labor, society, and politics operate. It is particularly useful in understanding Fordism because of its ability to capture the periodic growth and decline of national systems of capitalism, and link this growth and decline to the web of national social and political institutions and cultural attributes within which capitalist production is embedded.

Fordist production is an answer to an historically specific economic circumstance that occurred in the 1920s and in the immediate postwar period: how to organize production within a set of national economies (increasingly linked internationally) that can absorb virtually all that is produced. Under this circumstance, the central problems of production are those of organizing supply, not of creating and organizing demand. The central attributes of Fordist production, mass production, industrial unionism and competitive bargaining, and vertical integration are best viewed as ways of ensuring adequate supply to meet demand. Price competition is perhaps the only logical form of competition within this type of production system.

The attributes of Fordist production and Fordism are extremely familiar to most inhabitants of the advanced capitalist nations. Given large and rapidly expanding mass markets that enable mass production, firms are able to harness huge internal economies of scale that allow correspondingly large investments in fixed-process technologies. Mass production methods similarly engender both the breaking-down of the product into ever finer parts, and hence a reliance on interchangeable parts and subassemblies (that may also be mass produced), and the corresponding breaking-down of work tasks leading to an extreme, Taylorist division of labor and a de-skilling of the labor force (Braverman, 1974). This latter phenomenon may be viewed as a sort of "mass production" of labor skills. At the same time technical, professional, and administrative professions emerged within manufacturing as the result of these same Taylorist principles of scientific management in the factory.

Because the central problem of production was how to meet demand, and because firms in the immediate postwar period functioned in an environment in which high, stable profits were the norm, they were willing and able to pay high, stable wages; improve working conditions; and stabilize employment in exchange for a reliable, organized workforce and control of the factory (Piore and Sabel, 1984, p. 64). Industrial unionism, typically organized along trade lines, both articulated labor demands to firms and, once forged, enforced labor agreements among workers. This social contract between capital and labor allowed for steady increases in the real salaries and wages of those employed in manufacturing within the advanced capitalist nations giving rise to a large and relatively affluent middle class within these societies. This not only served to narrow inequality, but it (and this is the true genius of Henry Ford's $5 day) created an ever-expanding pool of consumers for the products of industrial capitalism.

Since the chief problem in the immediate postwar period was securing adequate inputs to production in the face of large and highly standardized markets, Fordist firms were traditionally highly vertically integrated. At the interfirm level, the organization of an industry in a region has been one in which a few dominant firms assembling finished products are fed subcomponents and subassemblies from sources elsewhere in the region, nationally, or globally. Relationships within one of these "hub-and-spoke systems" might best be described as characterized by mutual distrust. The tendency has been to keep suppliers and buyers alike at arms length. Within the US, this "strictly business" relationship was sanctioned within society generally and given legal force by stringent antitrust regulations dating from the American experience with "robber baron" capitalism in the late 1800s.

Fueled by increasing tax revenues (which in turn were made possible by rising real personal and corporate incomes), both the absolute size of government and the number of government programs increased dramatically during the Fordist period. The larger size of government as well as the influence of scientific management philosophies led to a governmental structure that was increasingly internally compartmentalized. The period witnessed not only the growth in the size of government and its programs but an increasing complexity in governmental bureaucracy. In most advanced capitalist countries, traditional defense, infrastructural, and educational initiatives continued to consume the lion's share of the federal budget, while these same countries also added new programs (and corresponding expenditures) aimed at guaranteeing the social welfare rights of their citizens, and spreading the wealth more evenly. Social welfare expenditures grew especially dramatically in Europe. Among traditional categories of expenditure, important shifts occurred as well. For example, the growth of technical, professional, and administrative positions within manufacturing necessitated expansion of national university systems dedicated to the training of students studying extremely narrow subjects that were then translated into correspondingly narrow occupations within the white collar workforce.

Spatially, the heart of Fordist production was in the United States. Within the U.S., the Northeast and Midwest regions contained particularly high proportions of manufacturing. Yet all of the advanced capitalist nations have regions in which substantial Fordist manufacturing existed or exists. These include: in Great Britain, the Midlands, Glasgow, and southern Wales; in Germany, the Ruhr; in France, the Paris Basin and the regions of Alsace and Lorraine; in Italy, the Po Valley; and in Japan, the Tokyo and Osaka basins.

A NEW METHOD OF PRODUCTION?

By the late 1960s certain aspects of industrial production were undergoing rapid transition. For example, the "economic miracle" experienced by Italy in the early 1960s was associated with the export of goods, particularly furniture and textiles, distinguished by high quality and flawless design. Similarly, many of the current "lean production" methods used in the automobile industry today were first implemented in Japan during the 1960s. However, it was the global recession of the early 1980s (given our previous discussion, it should be noted that this recession affected different nations differently and at different times), that signaled the beginning of the current transition in industrial capitalism. The recession exacerbated existing overcapacity in several key and traditionally Fordist sectors globally, especially in the automobile and related metals industries. In an effort to secure new markets in the face of limited demand during the recession, firms actively sought to differentiate their products. This drastically reduced the size of production runs and effectively destroyed much of the economic rationale of Fordism.

There exists a divergence of opinion over whether the current period of industrialization marks an identifiable period of capitalism distinct from the past, or whether we are simply living in a period of "neo-Fordism." This new period of industrialization is typically portrayed as being characterized by a global trend toward a type of manufacturing that utilizes computer-controlled machinery in conjunction with labor possessing diverse skills embedded in a complex network of competing and cooperating firms to produce highly specialized products. While Piore and Sabel (1984), Scott (1988), Storper and Walker (1989), and Harvey (1989), among numerous others, have argued that the current period represents a distinct period of "flexible accumulation," as compared with previous "Fordist accumulation," others have stressed that the period of mass production is far from over (for example, Dohse, Jurgens, and Malsch, 1985).

Among those that largely agree that the current period represents a qualitatively different kind of manufacturing, disagreement exists over the precise definition of "flexibility." Gertler (1988, 1989), and Bessant and Haywood (1988) tend to narrowly define flexibility as the production of highly specialized products reliant on the combined use of microelectronics and engineering. Alternatively, Schoenberger (1988, 1989) and Gertler (1992) have taken a larger view of flexible manufacturing, one that incorporates not only new technology, but a reorganization of labor process, of firm management, and of the linkages that firms maintain with suppliers and customers. Others (notably Scott, 1988; and Florida and Kenney 1990), following the lead of Piore and Sabel (1984), have concerned themselves with the nature of flexible production complexes. A third group (Amin and Robins, 1990; and Blim, 1990) have stressed the intense competition and potential for exploitation of labor within flexible production complexes. Others accept the notion of a qualitative change in the manner in which production takes place, while acknowledging that the exact form that flexibility takes from place to place is socially conditioned (Florida and Kenney, 1994; Gertler, 1992, 1993; Ettlinger, 1991; Jameson, 1991; Sayer, 1989).

While the degree of difference between industrial capitalism as practiced in the immediate postwar period and as practiced today remains a topic of heated debate, it is clear that conditions have changed. While Fordist production is one answer to organizing production within a set of national economies (increasingly linked internationally) that can absorb virtually all that is produced, the move toward more flexible approaches to production is the response to a need by firms to create and organize demand in an age of overcapacity and global competition, and the current economic transition must be understood against these changed economic circumstances. The central attributes of current industrial capitalism, small batch production, labor contingency, and vertical disintegration are best viewed as ways of creating highly customized products for niche markets within an uncertain and risky business environment.

The basis for competition is also both different and broader than under Fordism which relied primarily on price competition afforded by ever-increasing economies of scale. In the current environment, competition rests on external economies of scope and economies of scale that must be maintained in the face of highly differentiated product lines. This imperative engenders an ever decreasing technical division of labor. Price competition is still important to competition, but only within the overarching context of flexibility and quality. The flexible firm thus competes along dimensions that not only include traditional Fordist concerns such as price, but on quality, degree of customization, and timeliness of delivery as well.

The changed context of capitalist competition has altered the use of technology, labor, the way in which firms are organized internally, and the way in which they interact. Technologically, the introduction of new "flexible manufacturing" techniques have increased the ability of firms to respond to the very particular and frequently changing demands of niche markets. Flexible manufacturing is most often associated with the use of sophisticated computer controlled machine tools that simultaneously allow high-speed production, rapid change-over between product lines, and high quality. New information technologies have allowed the integration and coordination of activities both within and across firms and this increased integration and coordination has led to phenomenon such as "just-in-time" production.

The introduction of flexible machine tool technology has often gone hand-in-hand with alterations in the "society" of work. The increased horizontal authority of existing managers and worker empowerment are two of the more visible aspects of this alteration. These changes in the "society" of work have led to dramatic reductions in both the number of production workers (the recession of 1979-83) and the number of middle-level managers (the recession of 1990-91). They have also led to an increasing contingency of the labor force, as firms seek to gain increased labor flexibility by altering the number of workers (numerical flexibility), the wage of those workers (wage flexibility), or the tasks that workers perform (functional flexibility).

Perhaps one of the areas most strongly affected by the recent transition is the area of interfirm relations. While the networks of competing and cooperating firms discussed by Piore and Sabel (1984, p. 29) have so far failed to occur with any sort of regularity, interfirm relations are quite different from those that prevailed 30 years ago. What is most striking is that there has occurred a surprising variety of new interfirm relationships. These have been variously classified as Italianate, neo-Marshallian, tiered, and hub-and-spoke (Markusen, 1994). These forms span the spectrum from that anticipated by Piore and Sabel (Italianate and neo-Marshallian) to those relatively similar to those that prevailed under Fordism. This wide variety of forms stems, on the one hand, from the varying uncertainty surrounding demand among firms and, on the other, from the varying importance of the minimization of transaction costs through long-term cooperative relationships among firms. The first dictates the firm-specific level of vertical disintegration under uncertainty (since the vertical integration of multiple steps of production makes sense only within an economic environment of certainty of demand), while the later suggests that, although it is still economically rationale for firms to remain disintegrated, cooperative behavior decreases transactions costs below that which would prevail in the absence of such cooperative agreements for some firms. Thus cooperation may have a real economic payoff in terms of competitive advantage, but these payoffs are both industry and firm-specific. Thus, it is not uncommon to see small, vertically disintegrated but highly networked firms operating in the shadow of large, vertically integrated producers.

Just as there exists a geography of Fordist production, there is an emerging geography of post-Fordist production. As anticipated by regulation theory, the geography of post-Fordist production is different from that of the Fordist era. Within the U.S. the centers of post-Fordist production have been Route 128 in Massachusetts, and Silicon Valley and the Los Angeles basin in California. While the traditional manufacturing core of the U.S .has recently undergone somewhat of an economic revival, the exact nature of that revival is open to question (Knudsen, 1996). In Great Britain, post-Fordist production is concentrated along the M5 between London and Oxford, while the traditional industrial heartlands of Britain, particularly the Midlands, have languished. In Germany, BadenWürttemburg rivals the Ruhr as the industrial heartland, while in France, Toulousse is a rising industrial star. In Japan, Aichi province and Tokyo continue their prominence, and in Italy, Emilia-Romagna is the center of production. Other counties in Europe have experienced geographic shifts as well. For example, the industrial heart of Denmark in now western Jutland, not metropolitan Copenhagen. Further, while in some cases these shifts are associated with the emergence of new industries (electronics in the case of Great Britain and the U.S., plastics in the case of Denmark, airplanes in the case of Toulousse), as often as not they involve a radically different approach to the production of a traditional commodity (textiles in Italy, automobiles in Japan, machine tools in Germany).

IN THE SEQUEL

The remainder of this volume stresses five dimensions of the current transition in industrial capitalism: technology, labor relations, firm organization, public policy, and geography. Naturally, it is impossible within the context of an edited volume to provide a complete synthesis of all of these dimensions. Instead, the collection of essays is topical. Edward Malecki thoroughly discusses information technologies and their implications for flexibility and competitiveness. The essays by Kenney and Curry, and by Mikulecky reflect two aspects of the emerging labor relation: the bifurcation of the labor market, and the emerging skills crisis. The essays by Pavlik, Warf, and Bosworth reflect three critical aspects of firm strategy: labor strategy, the dualistic role of services and manufacturing in modern society, and the increasing importance of networks and strategic alliances in the global economy. In the latter portion of his essay, Bosworth examines the role of public policy in promoting networks and alliances. The role of public policy is also the topic of the essay by Wheeler. The essays by Leigh and Cox discuss various aspects of the geography of flexible production. The essay by Leigh concerns the changing needs of capital as it affects the urban built environment, while Cox discusses in more general terms the role of flexible accumulation in the transformation of space. Each of the essays is discussed in greater detail below.

In the next chapter Edward Malecki discusses a triad of concepts: technology, flexibility, and competitiveness. Of these, he argues that technology is the principal force behind capitalism. Traditionally, technology has been construed as meaning "machines for production." While technology does include machinery, and the emerging period of flexible manufacturing utilizes sophisticated computer integrated machine tools, management technologies and organizational technologies exist as well. This paper both defines and explores the various kinds of machine, management, and organizational technologies utilized in advanced capitalist nations.

Martin Kenney, in Chapter 3, examines the sharp bifurcation that exists today in the ways that labor is utilized in the production of goods and services. Increasingly, labor can be divided into that labor which is highly skilled and is at work in the modern computerized factory or service center, and that labor which has low skill, but that nevertheless can be exploited cheaply and flexibly in "sweatshop" environments. He compares the utilization of labor in Japanese factories, American factories, and factories in the Mexican maquiladoras.

Larry Mikulecky examines the new requirements that modern workplaces place on workers and on public education. A sizable minority of the U.S. workforce is consistently unable to perform many of the tasks that are required in the modern workplace. Further, despite repeated attempts, there appears to be little improvement in the skill level of recent high school graduates. A number of new initiatives aimed at smoothing the school-to-work transition offer hope, but have so far affected very few students.

Claire Pavlik notes that in response to the crisis of the 1980s, firms have pursued divergent strategies with regard to labor process. There appears to be no consensus among firms on what strategies work and what strategies don't work. While some firms are presently engaged in strategies that primarily are extensions of Fordist practices in which there as yet does not appear to be any "clear break from the past," others are attempting to revolutionize labor process. As yet, no one "best practice" model has appeared.

Barney Warf's contribution discusses the role of services in the industrial transition. The services are absent from much of the dialogue concerning the recent transformation of advanced capitalist economies. This exclusion, which stems from the traditional viewpoint that services are nonbasic industries, constitutes a serious oversight if we wish to understand the modern economy in which services play a critical, value-adding role. This critical role stems from five factors. First, the vertical disintegration of large manufacturing firms has led to the increased purchase of services outside the firm. Second, information technology has led to the formation of information hubs within the global economy. Within these "command and control" centers a post-industrial economy flourishes by providing services to industrial centers across the globe. Third, since demand for services is income elastic, rising incomes in the industrialized world have led to a rise in services. Fourth, aging populations in the industrialized world have created a rise in health care services. Finally, as business becomes increasingly global, a larger proportion of any firm's employment is dedicated to nondirect production such as legal services and marketing.

An important aspect of competitiveness today is the ability to wrap the commodity for sale within a large bundle of ancillary services, according to Brian Bosworth. Small firms particularly have a difficult time of this because their small size inhibits their ability to provide this bundle. An increasingly common strategy employed by small firms through which they can increase delivery of ancillary services to their customers is networks and alliances. While networks and alliances offer a number of advantages, managerial inertia and poor public policy have hampered their widespread adoption.

In Chapter 8, James Wheeler discusses the policy implications of flexibility. Much of the literature on flexibility is bound up in a discourse on definitions. The demands on policymakers for flexibility by constituents is great, but are not always in the state's or locality's best interests. Furthermore, states are organized as hierarchial organizations. The privatization and reinventing government movements are important components of a new, flexible, and rejuvenated state.

Much has been made of the "new industrial spaces" occupied by flexible manufacturing, especially in the neo-Marshallian districts of Los Angeles, Baden-Württemburg, and the Third Italy. Fordism, as it collapsed, left in its wake considerable infrastructure and real estate in the urban areas of the Midwest and Northeast, according to Nancey Green Leigh. Communities have struggled to renovate these "brownfield" sites so that they can be redeveloped. This chapter traces the history of two brownfield redevelopments in Milwaukee, Wisconsin.

The demise of Fordism and the emergence of flexibility should not be taken as a given, according to Kevin Cox. Economic transformation is contingent on the local social and political milieu, just as those milieu are contingent on the economy. Chapter 10 provides a Realist reinterpretation of recent events in the advanced capitalist nations. This reinterpretation is then used to re-examine the capitalist dynamic.

REFERENCES

Aglietta, M. (1979). *A Theory of Capitalist Regulation*. (English ed. 1979.) London: New Left Books.

Amin, A. and Robins, K. (1990). "Flexible Specialization and Small Firms in Italy: Myths and Realities." In F. Pyke, G. Becattini, and W. Sengenberger (Eds.), *Industrial Districts and Interfirm Cooperation in Italy*. Geneva: ILO.

Berry, B.J.L. (1991). *Long-Wave Rhythms in Economic Development and Political Behavior*. Baltimore: Johns Hopkins University Press.

Bessant, J. and Haywood, B. (1988). "Islands, Archipelagos and Continents: Progress on the Road to ComputerIntegrated Manufacturing," *Research Policy* 17: pp. 349–62.

Blim, M. (1990). *Made in Italy: Small-Scale Industrialization and Its Consequences*. New York: Praeger.

Boyer, R. (1986). *The Theory of Regulation: A Critical Analysis*. English translation 1988. New York: Columbia Press.

Boyer, R. (1988). "Technical Change and the Theory of Regulation." In G. Dosi et al. (Eds.), *Technical Change and Economic Theory*. London: Pinter, pp. 67–94.

_____ (1990). *The Regulation School: A Critical Introduction.* New York: Columbia University Press.

Braverman, M. (1974). *Labor and Monopoly Capital.* New York: Monthly Review Press.

Clark, G.L. (1989). *Unions and Communities Under Siege*. Cambridge: Cambridge University Press.

Dohse, K.; Jurgens, U.; and Malsch, T. (1985). "From 'Fordism' to 'Toyotism': The Social Organization of the Labor Process in the Japanese Automobile Industry," *Politics and Society* 14: pp.115–46.

Dunford, M. (1990). "Theories of Regulation," *Environment and Planning D: Society and Space* 8(3): pp. 297–321.

Ettlinger, N. (1991). "Antecedent Conditions and the Roots of Competitive Advantage in California and Japan," *Annals, Association of American Geographers* 81: pp. 391–407.

Florida, R. and Kenney, M. (1990). "High Tech Restructuring in the USA and Japan," *Environment and Planning A* 22: pp. 233–52

_____ (1994). "Institutions and Economic Transformation: The Case of Postwar Japanese Capitalism," *Growth and Change* 25(2): pp. 247–62.

Gertler, M. (1988). "The Limits to Flexibility: Comments on a Post-Fordist Vision of Production and its Geography," *Transactions, Institute of British Geographers N.S.* 13: pp. 419–32.

_____ (1989). "Resurrecting Flexibility? A Response to Schoenberger," *Transactions, Institute of British Geographers N.S.* 14: pp. 109–12.

_____ (1992). "Flexibility Revisited: Districts, Nation-States, and the Forces of Production," *Transactions, Institute of British Geographers N.S.* 17: pp.259–78.

_____ (1993). "Implementing Advanced Manufacturing Technologies in Mature Industrial Regions: Towards a Social Model of Technology Production," *Regional Studies* 27(7): pp. 665–80.

Goodwin, M.; Duncan, S.; and Halford, S. (1993). "Regulation Theory, the Local State, and the Transition in Urban Politics," *Environment and Planning D: Society and Space* 11(1): pp. 67–8.

Gruber, A. and Nowotny, H. (1990). "Towards the Fifth Kondratiev Upswing," *International Journal of Technology Management* 5: pp. 431–71.

Hall, P. (1985). "The Geography of the Fifth Kondratieff." In P. Hall and A. Markusen (Eds.), *Silicon Landscapes*. Boston: Allen and Unwin, pp. 1–19.

Harvey, D. (1989). *The Condition of Post Modernity*. Oxford: Basil Blackwell.

Jameson, F. (1991). *Postmodernism, or, The Cultural Logic of Late Capitalism*. Durham, NC: Duke University Press.

Jessop, B. (1990). "Regulation Theories in Retrospect and Prospect," *Economy and Society* 19(2): pp. 153-217.

Jessop, B.; Nielsen, K.; and Pedersen, O.K. (1993). "Structural Competitiveness and Strategic Capacities: Rethinking the State and International Capital." In S. Sjøstrand (Ed.), *Institutional Change: Theory and Empirical Findings*. New York: M.E. Sharpe, pp. 227–62.

Kenney, M. and Florida, R. (1989). "Japan's Role in a Post-Fordist Age," *Futures* 21(3): pp. 136–51.

Knudsen, D.C. (1996). *Regional Economic Restructuring: A Case Study of the American Midwest, 1950–1992*. Newberry Park CA: Sage Publications.

Lipietz, A. (1986). "New Tendencies in the International Division of Labor." In A. Scott and M. Storper (Eds.), *Production, Work, Territory*. Winchester: Allen and Unwin, pp. 16–40.

Mandel, E. (1980). *Long Waves in Capitalist Development*. London: Cambridge University Press.

Markusen, A.R. (1994). *Sticky Places in Slippery Space: The Political Economy of Postwar Fast-Growth Regions*. New Brunswick NJ: Project on Regional and Industrial Economics, Rutgers, the State University of New Jersey.

Marx, K. (1977). *Capital-Volume 1*. (English trans.) New York: Vintage Books.

Massey, D. (1984). *Spatial Divisions of Labor: Social Structures and the Geography of Production*. London: Macmillan.

Peet, R. (1991). *Global Capitalism: Theories of Societal Development*. London and New York: Routledge.

Piore, M. and Sabel, C. (1984). *The Second Industrial Divide*. New York: Basic Books.

Rothwell, R. (1982). "The Role of Technology in Industrial Change: Implications for Regional Policy," *Regional Studies* 16: pp. 361–69.

Sayer, A. (1989). "Post-Fordism in Question," *International Journal of Urban and Regional Research* 13(4): pp. 666–95.

Schoenberger, E. (1988). "From Fordism to Flexible Accumulation: Technology, Competitive Strategies and International Location," *Environment and Planning D: Society and Space* 6: pp. 245–62.

_____ . (1989). "Thinking About Flexibility: A Response to Gertler," *Transactions, Institute of British Geographers N.S.* 14: pp. 98–108.

Scott, A. (1988). *New Industrial Spaces*. London: Pion.

Storper, M. and Scott, A. (1989). "The Geographical Foundations and Social Regulation of Flexible Production Complexes." In M. Dear and J. Wolch (Eds.), *Power of Geography*. Boston: Unwin Hyman, pp. 21–40.

Storper, M. and Walker, R. (1989). *The Capitalist Imperative*. New York: Basil Blackwell.

Teague, P. (1990). "The Political Economy of the Regulation School and the Flexible Specialization Scenario," *Journal of Economic Studies* 17(5): pp. 32–54.

Tickell, A. and Peck, J. (1992). "Accumulation, Regulation and the Geographies of Post-Fordism: Missing Links in Regulationist Research," *Progress in Human Geography* 16(2): pp. 190–218.

2

TECHNOLOGY, COMPETITIVENESS, AND FLEXIBILITY: CONSTANTLY EVOLVING CONCEPTS

Edward J. Malecki
Department of Geography, University of Florida, Gainesville, FL 32601 USA

INTRODUCTION

This chapter examines three concepts central to the changes evident in the world economy, and their impacts on regions and places, firms, and people. Technology is perhaps the most important of the three because it has the broadest scope and thus directly affects the other two concepts: flexibility and competitiveness. Technology also is the one least able to be controlled by individual firms, since technology originates in other firms, especially competitors, and in research institutes and universities throughout the world. However, technology provides only a portion of the advantages described as flexibility or as competitiveness. All three are central phenomena of the 1990s, where global capitalist competition has pushed small firms to be international and seek out export markets, and pushed large, transnational firms to form networks of partners, large and small, that allow them to become flexible.

The three concepts are addressed in turn, followed by a brief consideration of flexible places. Each topic is illustrated with reference to a large (and growing) literature. The relative significance of flexibility, for example, may be overstated compared to that of technology and innovation.

TECHNOLOGY

Technology refers to the general capabilities of human society to transform nature into useful products for human consumption. This takes four main forms:

- general theoretical and practical understanding about how to do things (social knowledge),
- objects (goods),
- installed techniques of production (processes), and
- the personal know-how and abilities of workers (skills).

Technological change occurs in all these forms or dimensions (Storper and Walker, 1989, pp. 50–1). An interpretation of this multidimensional capability centers around knowledge, seen explicitly in social knowledge and skills, and implicitly in goods and processes. Moreover, there are interactions among all four forms, such that we cannot separate them completely.

The result of the increased intensity of technological change is that, in contrast to the linear innovation model underlying, for example, product cycle theory, there is no longer a clear-cut differentiation and no clear sequence from research to development and to production and distribution. Instead, the innovation process is highly interdependent, and includes a variety of ways of learning and of generating new knowledge (Kline and Rosenberg, 1986; Malecki, 1991). For example, by learning from customers and from suppliers, by interacting with other firms, by searching for new technologies, and by taking advantage of spillovers from other industries, firms are able to enhance their technological capabilities (Bell, 1984; Lundvall, 1988; Malerba, 1992; David, 1993). However, these possibilities for learning technology also present the challenge that no firm is able to monitor all information sources completely. Technological learning includes three types: search (seeking new knowledge, either internally or externally), accumulation and diffusion (dissemination inside the firm so as to prevent repetition of search activities), and review of previous knowledge (Dodgson, 1992).

As the number of product varieties increases, producers of goods and services must make use of both existing and new capital equipment (processes) as well as the new organizational technologies (or knowledge) which lie behind each of the possible configurations of people and equipment. Their organizational knowledge develops through the "coevolution of organization and technology" that makes technology flexible (and dynamic) rather than static (Dosi and Kogut, 1993). Jaikumar (1986, p. 70) observed that "It is not how the equipment is used that is important. Success comes from achieving continuous process improvement through organizational learning and experimentation."

The importance of this co-evolution and continuous learning should not be underestimated. Part of the reason concerns the adoption of new technology, a process we now know to be far more complex than had been thought previously. For example, new machinery not only must be purchased, it must become embedded in the firm and coordinated with all its other machinery, processes, logistics, and production practices.

This implementation of new technology means that technology is acquired only through long-term commitment, involving learning and strategic actions to absorb new skills (Prahalad and Hamel, 1990). One way in which this frequently occurs in the adoption of new capital equipment is through frequent interaction between the purchasing firm and its equipment supplier. Gertler (1993) makes the case that this interaction occurs best when manufacturers and customers are located in close proximity, because this facilitates frequent interaction and ensures that the specific needs of the adopting firm are met.

The purposeful integration of technology into the essence of the enterprise has taken place perhaps most in Japanese firms. "Technology fusion," as Kodama (1992) calls it, comprises a number of mechanisms by which workers and managers become thoroughly familiar with diverse technologies in order to obtain innovative new technologies. This fusion is made possible by intense interaction and communication, job rotation, and organizational learning (Bowonder and Miyake, 1993). The importance of technology fusion or integration has spread beyond Japan. A widely cited article by Prahalad and Hamel (1990) stresses that each firm develop its core competence: " . . . the collective learning in the organization, especially how to coordinate diverse production skills and integrate multiple streams of technologies" (p. 82). The challenge is to develop core competencies broadly and in the direction of change and progress, rather than only along existing trajectories (Utterback, 1994, pp. 217–22). This requires that firms be prepared for future technologies, not only present technologies (D'Aveni, 1994; Hamel and Prahalad, 1994).

COMPETITIVENESS

The imperative to be competitive stems from the increased globalization of competition as manufacturers face new competitors—especially from Asia, Mexico, and elsewhere—able to produce products that meet world market standards of price, delivery, service, and quality. No sector is immune from the technologically derived standards of quality and price that are set by global, world-class firms. "Competition itself is undergoing far-reaching changes. Whereas in the past it was predominantly price-driven, it has now become technology-driven" (Stevens and Andrieu, 1991, p. 116).

To attain competitiveness, whether called productivity, efficiency, or product quality, requires capital, skilled labor, training to enhance labor skills, and information about markets (Hitchens, Birnie, and Wagner, 1992). This information and knowledge comprises a number of specific entrepreneurial competencies, specific knowledge that must be accumulated within the firm. Johannisson (1990) has described them simply as know-how, know-what, know-when, know-who, and know-why, and one could add know-where to this list. When this diverse knowledge and these capabilities are in place, firms are able to shift from competitive strategy (staking out a position in a stable environment) to strategic flexibility, which includes being able to rapidly switch gears from product development to production (Hayes and Pisano, 1994).

Large firms develop a wide network of contacts and information sources on markets, suppliers, and technology. Gaining access to outside sources of knowledge is a principal impetus for strategic alliances and other forms of collaboration. Consequently, many experts, among them Dicken (1992) and Harrison (1994), believe that these large-firm networks give global firms advantages which small firms cannot match. Small firms find the demands of the market and the lack of resources available to them keeps them less than fully informed about markets and demands elsewhere. The complexity of industrial organization, especially regarding supplies of inputs and distribution of outputs, poses great challenges to small enterprises, which often struggle to maintain competitiveness as innovation increases in importance (O'Farrell and Hitchens, 1989). The knowledge needed to compete comes most often from suppliers and customers, and from other firms in the locality.

Market information, especially about export markets, is particularly difficult for many small firms to obtain. Trade publications, trade shows, and directories provide formal sources of information. Informal information-gathering can be initiated at trade shows, but works largely through a contact network of individuals. These contact networks, in the right setting, also provide avenues to sources of financing and investment support (Birley, 1985). Knowledge of international markets requires both greater resources and greater management capacity, both of which may be lacking in small firms (Christensen, 1991).

The disadvantages of small size in information-gathering especially affect firms in places where the local networks are inadequate for firm success. MacPherson's recent (1991) work in the Buffalo, New York area shows that it is not necessarily the local economic environment of such a declining region that leads to a lack of competitiveness, since technical information and other services can be imported. However, a striking gap appears between large firms, which generally import such information, and small firms, which by and large do not. Generally, more innovative firms are those with a high outward orientation and widespread networks of contacts (MacPherson, 1988; Rothwell, 1991).

Many small firms are able to be innovative, even if undertaking little or no research and development (R&D), by relying heavily on their external networks for information. Innovative small firms tend to have dense networks of connections with other small firms, in addition to links with large firms (Rothwell, 1991). It must be pointed out that R&D is not the key input or activity to measure. R&D figures will not include network activities, learning, informal R&D, and other means of enhancing a firm's (and its employees') knowledge (Malecki, 1991; Møller, 1991). Møller, for example, attributes Denmark's ability to excel in low-technology exports to these knowledge-enhancing activities. The innovativeness supported by local inter-firm networks not only supports existing firms; it also presents opportunities for entrepreneurs to start new businesses to serve newly identified markets. The importance of networks, of market information, and of innovative niches sparks innovation in both high-technology industries and in traditional sectors (Doeringer, Tekla, and Topakian, 1987; Miller, Brown, and Centner, 1987; Young and Francis, 1989). Miller, Brown, and Centner, (1987) provide evidence that handicraft firms in the U.S. South (often just individual proprietors) vary greatly in competitiveness according to several "network" factors. Competitiveness in craft sectors is largely an issue of new product generation to replace old products for which cheaper imitations have become available from Asia, Mexico, and other low-cost production areas. To a lesser extent, competitiveness also involves issues of quality, which depends on production techniques. The network factors include education and training (which broadens the individual's network of contacts and information sources), the sources of information utilized for new product ideas (for example, personal contact networks, trade publications, short courses), and the extent to which one participates in trade shows, where contact with customers, suppliers, and competitors is facilitated.

A great deal of information flow among firms depends on "gatekeepers" to external networks and on informal, unplanned interaction. Especially important is information that is oriented toward national and international markets. Keeping up with industry best practice internationally is a key concern of all firms, and determines their competitiveness in a global economy (O'Farrell and Hitchens, 1988, 1989). Similarly, a lead firm may play the role of coordinator within a "constellation" of firms, providing a link to sources of information, innovation and technology (Shaw, 1991).

The issue of competitiveness, then, is complex, and largely one of knowledge: knowledge about the needs of customers (old and new) and about technologies which will enable those needs to be met. As customer needs have become less standardized, the need for nonstandard, flexible, responses has grown.

FLEXIBILITY

"Flexibility is the ability to change or react with little penalty in time, effort, cost or performance" (Upton 1994, p. 73). Flexibility represents the response of firms to a changed set of rules that govern competition. Instead of standardized, one-product-fits-all, high-volume production, market niches, "mass customization," and the goals of product innovation and quality require that market information and production be well-coordinated (Goldhar, 1986; Pine, 1993). In order to address the decline in mass markets, more flexible forms of production, of interfirm relations, and of work are forcing changes into the orderly sequence suggested by the product cycle (Ayres and Steger, 1985; Goldhar, 1986).

The emergence of flexibility has had deep effects, often characterized as a move from the Fordist towards a post- (or neo-) Fordist paradigm (Harvey, 1989; Benko and Dunford, 1991; Storper and Scott, 1992). Consumer demand is more differentiated, product cycles have become shorter, and the pace of technological change has quickened. This generally has increased both the need for firms to gain flexibility in their operations and also the need for continual innovation. The relative growth of small firms is an effect of this trend to a certain extent, as is the restructuring of large firms and their move towards "lean production" (Malecki and Todtling, 1994).

The characteristics of the new flexible economy are complex, involving capital-intensive investment, labor flexibility (for both skilled and unskilled workers), and new ways of organizing production and interactions with suppliers and customers. The principal reason for flexibility is the speeding up of product cycles, which means that economies of scale and the large production volumes that are assumed no longer apply, and that a much greater amount of attention must be paid to product innovation in order to generate the succession of new product cycles. It is useful to distinguish among these three types of flexibility, for they represent truly different forms of organizing production, directed toward the twin goals of product innovation and quality:

1. production flexibility: the adoption of systemic computer-integrated manufacturing technologies and forms of production organization;
2. flexibility of inter-firm relations: the development of new systematic relationships with other plants and other firms;
3. labor flexibility: the adoption of a new flexible labor process in which the past tendencies towards the increasing division of tasks, the deskilling of work, and the removal of control over production from the worker are reversed (Hoffman and Kaplinsky, 1988, p. 331).

Flexibility in Production

Although production organization must incorporate flexibility, generally production flexibility focuses on the spectrum of changes which result from automation. Automation based on computers and microelectronics has three qualities which push it beyond the concept of other labor-saving capital investment. First, flexibility is built into the machinery, which is general-purpose rather than product specific. No set-up is required to switch to a different product. A new product can be built simply by programmed instructions on machines which can produce many different products (Shaiken, 1984, pp. 6-7). Second, this flexibility leads to a broader scope of application. The expense of new technology might be greater than of previous generations of machines, but it provides a great deal more flexibility (to use the word in yet another sense) in product variety. This form of flexibility is especially important, because a firm with such a capability is able to handle both routine, volume production and more difficult (and profitable) nonstandard orders, which allow it to take the lead on small-volume new product introduction (Hayes and Wheelwright, 1984, p. 40). Finally, an important goal in itself of programmable technology is improved quality and reliability, a factor more important than variety or price in many markets (Sciberras, 1986; Starr and Biloski, 1985).

The push toward flexible production within firms and the profusion of strategic alliances among firms can be understood through the concept of economies of scope (Charles, Monk,and Sciberras, 1989; Teece, 1980). Internal economies of scope accrue to firms which can produce two or more different, related products together more cheaply than in isolation, ideal for situations of product differentiation. External economies of scope result when firms can use their know-how to apply to the activities, particularly innovative activities, of other firms and other markets.

The Fordist priority on mass production based on economies of scale has given way, then, to economies of scope, cost savings based on producing a variety of products or services in small, even one-of-a-kind, batches. Ideally, such a factory "can produce a continuous stream of different product designs at the same cost as an equal stream of identical products" (Goldhar, 1986, p. 27). Such totally flexible production systems custom-tailor products to the needs and tastes of individual customers, combined with the power, precision and economy of modern production technology. The strategic objective is to deliver high-quality products tailored to each customer at mass-production prices (Dertouzos et al., 1989, p. 131). Thus, the mass customization of Pine (1993) and the lean production of Womack, Jones, and Roos (1990). The pressure to customize for all customers, rather than to sell duplicate products, has a cost in reducing response time. It takes more time to be innovative and to tailor custom products than to produce standardized products (McCutcheon, Raturi, and Meredith, 1994).

In order to retain some efficiencies and to keep production volumes high, "design for manufacture" simplifies products to minimize the number of parts and to increase the number of common parts in different final products (Charles, Monk, and Sciberras, 1989, pp. 135–7). Upton (1994, p. 74) lists 15 categories of manufacturing flexibility, including routing, product, mix, volume, long-term, short-term, expansion, machine, operation, and design-change flexibility.

Because automation technologies and their organizational demands are easily misunderstood and misused, and because of the number of categories and dimensions of flexibility, they are particular obstacles for small firms (Gros-Pietro and Rolfo, 1989; Upton, 1994; Warner, 1987). "Flexibility is much more an organizational property than a technical one" and is best managed by large firms (Bessant and Haywood, 1988, p. 359). The difficulty of integrating new technology keeps large firms clearly in control as flexible, even custom, production, once the province of small firms, prevails (Harrison, 1994).

Agile manufacturing is the latest term for this ultra-flexible firm (Goldman and Nagel, 1993). In this context, it is clear that technology does not drive the firm. It is the management of the firm that decides, on managerial grounds, the kinds of technical expertise, and the directions of R&D that will be supported internally, or acquired from other firms, in the form of personnel, patents, licensing agreements, or reverse engineering. It is, again, management that decides for each firm what technical knowledge and which technologies are to be used, what they are to be used for, and on what scale (Goldman and Nagel, 1993, p. 19).

Originating primarily in the U.S. Department of Defense, agile manufacturing integrates total quality management, concurrent engineering, and flexible manufacturing, with emerging trends such as intelligent processing, computer-aided test and inspection, and advanced systems integration (Ligus, 1990; *Manufacturing Engineering*, 1993; Sheridan, 1993; Youssef, 1992).

Flexibility in Inter-Firm Relations

Automated flexible production systems, in turn, require a tightly controlled stream of inputs tailored to the needs of production on both a short-term and a long-term basis. Some of these are best supplied within the firm itself, based on its core competencies. Others are better performed externally, via "strategic outsourcing" (Quinn and Hilmer, 1994). On a daily scale, supplies of inputs are best managed at low cost by employing the just-in-time (JIT) system of inventory management, developed by Toyota during the

1950s. This system demands that suppliers provide a higher level of quality, defined as fewer defective parts per order with a goal of zero defects. In short, parts producers must operate at a higher frequency ("more often") and in smaller volumes ("less") than was true under traditional production organization (Arnold and Bernard, 1989, p. 415; Hoffman and Kaplinsky, 1988; Lenz, 1989).

JIT affects many other activities of production, and this favors its adoption by large, rather than small, firms. It permits or necessitates, for example, integrated data processing for R&D, procurement, production planning, inventory control, and marketing (Arnold and Bernard, 1989). In this way, (JIT) relationships appear to promote larger, rather than smaller, firm sizes in suppliers. The number of suppliers tends to drop substantially under JIT, also promoting larger firm size, since a proliferation of input producers is unwieldy to manage, and single-sourcing becomes commonplace. Suppliers may face increased costs, among them for computer systems and communication links, largely related to a push for high quality, "zero defects" production (Arnold and Bernard, 1989). The JIT system leads to greater R&D and manufacturing system improvements at the factories of suppliers. This reliance on suppliers has the significant effect of reducing the number of suppliers to those few which are reliable in terms of quality, delivery, and response (Larsson, 1993).

A second form of flexibility in inter-firm relations is the formation of networks. Whereas the typical supplier-customer relationship is governed by market norms, networks fall somewhere in between markets and intra-firm hierarchies (Larsson, 1993; Thorelli, 1986). Large firms typically develop their networks through strategic alliances with other firms that offer complementary skills and knowledge, even, or especially, competitors (Chesnais, 1988; Cooke, 1988; Teece, 1986). Smaller firms may develop similar arrangements, at considerable cost and effort, or develop networks of a more informal nature (Malecki and Tootle, forthcoming).

Flexible Labor

The presence of fewer routine tasks and few long production runs leads to a demand for a "highly skilled, flexible, coordinated, and committed work force" (Walton and Susman, 1987). Flexible production requires that workers themselves be "flexible" rather than rigid in the definition of the tasks for which they can be called upon (Blackburn, Coombs, and Green, 1985).

The task fragmentation, functional specialization, mechanization, and assembly-line principles of Fordism are replaced with a social organization of production based on work teams, job rotation, learning by doing, flexible production, and integrated production complexes (Kenney and Florida, 1988). These are largely work practices that evolved in Japan and which involve large firms. In a variety of respects, the "Japanese model" of production stands in sharp contrast to the Fordist model, but it also involves a blend of large, powerful companies and use of tight relations with supplier firms and teamwork in production (Hiraoka, 1989; IILS, 1993). This teamwork, usually embodied in "quality circles," is among the most frequent aspects of the "Japanization" of industry. Gertler (1988, p. 426) points out that the greater skills demanded of labor in such flexible production systems results in greater reliance on, and less flexible relationships with labor.

In addition to qualitative flexibility demanded by automated production, firms also make use of numerical labor flexibility through the use of contingent workers, a phrase used to designate workers hired at less than the standard wage, for less than the standard workweek or workyear, or with less than standard fringe benefits (Belous, 1989). This group is particularly common in the service sector, where part-time work is more common, among "homeworkers" and, increasingly, in manufacturing among subcontractors (Christopherson 1989; Gringeri, 1994). Harrison (1994) terms this "the dark side of flexible production" (see also Hudson, 1989).

At the opposite extreme from contingent workers, labor skills and flexibility are demanded. Small-volume batch production systems require less labor than do massive mass-production facilities, but workers must be sufficiently educated and skilled to perform their jobs on an ever-changing mix of products. Flexible production and the prominent place of small firms forces labor training to the forefront of local policy (Lloyd, 1989). Flexible firms need skilled and flexible workers. This need not mean exploitation of labor. Rather, it implies the flexibility of nonroutine work associated with innovation, niche markets, and economies of scope (Doeringer, Terkla, and Topakian, 1987).

Technological change raises the required level of theoretical knowledge necessary to perform a job. High technology industries, in particular, demand understanding of complex capital equipment, technical documentation, and integration of complex technologies (Bartel and Lichtenberg, 1987). The need for worker skills varies over the life cycle of a product (Carnevale, Gainer, and Meltzer, 1988; Flynn, 1993). The fullest range of skills will be found in nonroutine jobs, but a number of these basic skills are required by routine work as well. A machinist, for instance, must have technical reading skills in order to interpret blueprints and use precision instruments. He or she must be able to compute, for example, from fractions to decimals. In addition, problemsolving is needed to select the appropriate tools and machines for a particular task, and to tinker with and make adaptations for custom jobs (Carnevale, Gainer, and Meltzer, 1988, p. 22; Doeringer, Terkla, and Topakian, 1987, p. 88). In a setting where custom products and short product life cycles are the norm, many work tasks are nonroutine, personal and interpersonal skills and adaptability are essential.

FLEXIBLE PLACES

Based on the experiences of "success stories" of some high-tech regions (Silicon Valley, Route 128, M4 Corridor) as well as a number of other dynamic regions (Third Italy, Baden Württemberg, West Jutland) localized networks of small firms have been viewed as a model for the new flexible economy (Piore and Sabel, 1984; Pyke, 1992; Storper, 1993; Zeitlin, 1992). They produce a variety of customized products in small volumes, cooperating not only in production but also in distribution and in technological development (Garofoli, 1991).

Competitive advantages in such districts and in "new industrial spaces" (Scott, 1988) are achieved by high quality, know-how or design intensity as well as by flexibility and customization, instead of by scale economies and low cost. The network supports the innovation process by spreading risks, combining resources, and by sharing know-how and experience. Innovations are in fact regarded as a product of the local milieu since "it is often the local environment which is, in effect, the entrepreneur and innovator, rather than the firm" (Aydalot and Keeble, 1988). Spatial proximity and territorial agglomeration are considered as important preconditions for the creation and the functioning of such an innovative milieu (Stohr, 1986; Camagni, 1991). This is because highly skilled personnel are interregionally not very mobile, direct personal and informal contacts are highly relevant, and synergy effects are expected from a common cultural, psychological and political background. Cooke and Morgan (1993) propose a set of "key elements" of a networked region, ready for the support of flexible firms:

- a thick layering of public and private industrial support institutions,
- high-grade labor-market intelligence and associated vocational training,
- rapid diffusion of technology transfer,
- a high degree of interfirm networking, and, above all,
- receptive firms well-disposed towards innovation.

The entrepreneurial region is perhaps the most important type of flexible region. In it, local and regional networks create a "milieu," defined as the socioeconomic environment of an area resulting from the interaction of firms, institutions, and labor (Aydalot, 1986; Aydalot and Keeble, 1988; Maillat, 1991). This interaction leads to a common way of perceiving economic and technical problems and of finding respective solutions. Industrial districts embody the interaction and dense network of linkages that comprise a local production system, usually around a single industry or a cluster of highly related industries. Indeed, the industrial district itself can be seen as a collective entrepreneur, with not only firms, but inter-firm associations, worker organizations, financial institutions, and governmental agencies also playing important roles (Best, 1990).

The district or milieu fulfills the task of informal knowledge transfer through mobile labor, information links, supplier and customer links, as well as through various forms of cooperation. Although large firms also are frequently engaged in networks, local networks are said to be most important for small firms, which are less able to maintain costly boundary-spanning functions (such as R&D) or distant cooperations (Camagni, 1991; Maillat, 1991). Education and training institutions play an important role, as do institutions engaged in firm formation and technology transfer such as incubators and innovation centers (Cooke and Morgan, 1993; Garofoli, 1991; Pyke, Becattini, and Sengenberger, 1990).

Networks of innovative small firms are to be found in various types of regions, leading to some hope that they might be a general model for regional economic development (Aydalot, 1986; Aydalot and Keeble, 1988; Cooke and Morgan, 1993). Some of them, particularly the high tech industries, cluster around large metropolitan areas (such as London, Milan, Paris, Madrid and Barcelona), others are in intermediate regions with more diffused industry (Third Italy: clothing, textiles, leather goods, ceramics, furniture; Baden Württemberg: machine-tool industry; West Jutland in Denmark: garment/knitting, furniture; and Andersdorp (Sweden): mechanical engineering). Finally there are also examples in "old industrialized" regions in the process of restructuring (the Swiss watch-producing region).

Nor do all spatial agglomerations of small firms in the same or related sectors necessarily comprise an industrial district or local industrial system. A functioning network or filiere of firms is essential for local development (Bull, Pitt, and Szarka, 1991). The web of local information flow, linkages, subcontracting, and sharing of new equipment, often eased by family ties, is the heart of the agglomeration advantages of industrial districts (Friedman, 1988, pp. 196–98). Cooperation permits the development of regional innovation complexes even in peripheral regions without agglomeration economies of the sort found in large urban regions (McArthur, 1989; Stohr, 1986). Based on work in Denmark, Hansen (1991) and Kristensen (1989) suggest that small-scale, specialized firms and a culture which supports entrepreneurship and cooperation are a new or different development model from the urban-based industrial archetype. In Jutland, far from metropolitan Copenhagen, for example, training and apprenticeships, artisan networks, and cooperatives are common.

The constitution of firm networks at the local and regional level is attributed not just to economic but also to sociocultural factors resulting from the history of the region (Garofoli, 1991). Local and regional authorities as well as chambers of commerce and industry in recent years have set up institutions to stimulate such local and regional networks. However, as can be seen from "success regions" like Baden Württemberg, Third Italy and others, networking has to be supported by a variety of institutions and actors and, in general, cannot easily be arranged by one or two public agencies. In addition to local and regional authorities and chambers, firms, educational institutions, universities, business services have to take their part. As in the case of the environment for entrepreneurship described above, most of the "ingredients" have to be already there in order to make it work (Smilor and Feeser, 1991).

Thus, we must be cautious about using localized small firm networks as a general model for regional development in the new flexible economy. At the conceptual level, a clear-cut transition from Fordism towards a networked small-firm economy has been questioned (Sayer, 1989; Amin and Robins, 1990). Large firms are far from leaving the stage (Martinelli and Schoenberger, 1991), and distant and global networks as well as supranational institutions gained importance in the 1980s (Dicken, 1992; Harrison, 1994). Gertler (1993) recently suggested that even local networks themselves are insufficient for long-term competitiveness of firms unless the local area also contains machine producers to support the local manufacturers.

Empirically, the evidence to date also is not entirely convincing (Malecki and Todtling, 1994). There are relatively few regions in Europe and the United States which have been serving as standard examples of dynamic small firm networks ("success stories"). Other regions have also been cited (Zeitlin, 1992), but the differences seem to be larger than the similarities. A closer look reveals that these regions are very heterogenous. They range from regions with family- and craft-based industries having sometimes very low levels of technology (especially in the districts of the "Third Italy") to sophisticated high-technology regions at the frontier of technological development. Consequently, there are many more regions which are linked in one way or another to large coordinating or leading firms, inside or outside the region (Schmitz, 1992) or which have no significant network links at all.

CONCLUSION

This chapter has shown that flexibility embodies technology as well as other organizational attributes. Indeed, flexibility, technology, and competitiveness are highly interrelated. Generally, firms with a larger number of sources of information about markets, about competitors, and about new techniques are those which introduce a greater number of new products or services. These sources may be mainly within the local base of firms, or they could be based on external networks (MacPherson, 1992). For all small firms that take competitiveness seriously, regular attendance at conferences and trade shows is a "must." No other single activity or information source is able to provide the diversity of contacts and types of information. Thus, an information network, whether local or nonlocal in scale, is critical. Outward-looking firms were overall more forward-looking as well, planning for new products, new techniques, new markets (Malecki and Veldhoen, 1993).

A related view is that of Kumpe and Bolwijn (1994), who suggest that the flexible firm has been superseded by the innovative firm. Firms which have already mastered quality and flexibility distinguish themselves from the competition by putting more innovation and technology into their products. The market demand for uniqueness can be met only by the combination of innovativeness, efficiency, quality, and flexibility (Kumpe and Bolwijn 1994, p. 42). The "bottom line" for all firms these days is a "hypercompetition" (D'Aveni, 1994) that demands that firms be multi-dimensional and multi-talented, while retaining strengths in their core competencies (Prahalad and Hamel, 1990). This is a tall order, but the alternative may be simply to fail to compete at all.

REFERENCES

Amin, A. and Robins, K. (1990). "The Re-emergence of Regional Economies? The Mythical Geography of Flexible Accumulation," *Environment and Planning D: Society and Space* 8: pp. 7–34.

Arnold, U. and Bernard, K.N. (1989). "Just-in-Time: Some Marketing Issues Raised by a Popular Concept in Production and Distribution," *Technovation* 9: pp. 401–31

Aydalot, P. (Ed.) (1986). *Milieux Innovateurs en Europe.* Paris: GREMI.

Aydalot, P. and Keeble, D. (Eds.) (1988). *High Technology Industry and Innovative Environments The European Experience.* London: Croom Helm.

Ayres, R.U. and Steger, W.A. (1985). "Rejuvenating the Life Cycle Concept," *Journal of Business Strategy* 6 (1): pp. 66–76.

Bartel, A.P. and Lichtenburg, F.R. (1987). "The Comparative Advantage of Educated Workers in Implementing New Technology," *Review of Economics and Statistics* 69: pp. 1–11.

Bell, M. (1984). "'Learning' and the Accumulation of Industrial Technological Capability in Developing Countries." In M. Fransman and K. King (Eds.), *Technological Capability in the Third World*. New York: St. Martin's Press, pp. 187–209.

Belous, R.S. (1989). *The Contingent Economy: The Growth of the Temporary, Part-time and Subcontracted Workforce*. Washington: National Planning Association.

Benko, G. and Dunford, M. (Eds.) (1991). *Industrial Change and Regional Development*. London: Belhaven.

Bessant, J. and Haywood, B. (1988). "Islands, Archipelagoes and Continents: Progress on the Road to Computer-Integrated Manufacturing," *Research Policy* 17: pp. 349–62.

Best, M.H. (1990). *The New Competition: Institutions of Industrial Restructuring*. Oxford: Polity Press.

Birley, S. (1985). "The Role of Networks in the Entrepreneurial Process," *Journal of Business Venturing* 1: pp. 107–17.

Blackburn, P.; Coombs, R.; and Green, K. (1985). *Technology, Economic Growth and the Labour Process*. London: Macmillan.

Bowonder, B. and Miyake, T. (1993). "Japanese Innovations in Advanced Technologies: An Analysis of Functional Integration," *International Journal of Technology Management* 8: pp. 135–56.

Bull, A.C.; Pitt, M.; and Szarka, J. (1991). "Small Firms and Industrial Districts, Structural Explanations of Small Firm Viability in Three Countries," *Entrepreneurship and Regional Development* 3: pp. 83–99.

Camagni, R. (1991). "Space, Networks and Technical Change: An Evolutionary Approach." In R. Camagni (Ed.), *Innovation Networks*. London: Belhaven, pp. 121–44.

Carnevale, A.P.; Gainer, L.J.; and Meltzer, A.S. (1988). *Workplace Basics: The Skills Employers Want*. Washington, DC: U.S. Department of Labor.

Charles, D.; Monk, P.; and Sciberras, E. (1989). *Technology and Competition in the International Telecommunications Industry*. London: Pinter.

Chesnais, F. (1988). "Technical Co-operation Agreements Between Firms," *Science Technology Industry Review* 4:51–119

Christensen, P.R. (1991). "The Small and Medium-Sized Exporters' Squeeze: Empirical Evidence and Model Reflections," *Entrepreneurship and Regional Development* 3: pp. 49–65.

Christopherson, S. (1989). "Flexibility in the US Service Economy and the Emerging Spatial Division of Labour," *Transactions of the Institute of British Geographers NS* 14: pp. 131–43.

Cooke, P. (1988). "Flexible Integration, Scope Economies, and Strategic Alliances: Social and Spatial Mediations," *Environment and Planning D: Society and Space* 6: pp. 281–300.

Cooke, P. and Morgan, K. (1993). "The Network Paradigm: New Departures in Cororate and Regional Development," *Environment and Planning D: Government and Policy* 11: pp. 543–64.

D'Aveni, R.A. (1994). *Hypercompetition: Managing the Dynamics of Strategic Maneuvering.* New York: Free Press.

David, P. (1993). "Knowledge, Property, and the System Dynamics of Technological Change." In *Proceedings of the World Bank Annual Conference on Development Economics 1992.* Washington: World Bank, pp. 215–48.

Dertouzos, M.L. et al. (1989). *Made in America.* Cambridge, MA: MIT Press.

Dicken, P. (1992). *Global Shift.* (2nd ed.) New York: Guilford.

Dodgson, M. (1992). "Strategy and Technological Learning: An Interdisciplinary Microstudy." In R. Coombs, P. Saviotti, and V. Walsh (Eds.), *Technological Change and Company Strategies.* New York: Academic Press, pp. 136–63.

Doeringer, P.B.; Terkla, D.G.; and Topakian, G.C. (1987). *Invisible Factors in Local Economic Development.* New York: Oxford University Press.

Dosi, G. and Kogut, B. (1993). "National Specificities and the Context of Change: The Coevolution of Organization and Technology." In B. Kogut (Ed.), *Country Competitiveness: Technology and the Organizing of Work.* New York: Oxford University Press, pp. 249–62.

Flynn, P.M. (1993). *Technology Life Cycles and Human Resources.* Lanham, MD: University Press of America.

Friedman, D. (1988). *The Misunderstood Miracle: Industrial Development and Political Change in Japan.* Ithaca, NY: Cornell University Press.

Garofoli, G. (1991). "Local Networks, Innovation and Policy in Italian Industrial Districts." In E. Bergman, G. Maier, and F. Todtling (Eds.), *Regions Reconsidered.* London: Mansell, pp. 119–40.

Gertler, M.S. (1988). "The Limits to Flexibility: Comments on the Post-Fordist Vision of Production and its Geography," *Transactions of the Institute of British Geographers NS* 13: pp. 419–32.

_____. (1993). "Implementing Advanced Manufacturing Technologies in Mature Industrial Regions: Towards a Social Model of Technology Production," *Regional Studies* 27: pp. 665–80.

Goldhar, J. (1986). "In the Factory of the Future, Innovation is Productivity," *Research Management* 29 (2): pp. 26–33.

Goldman, S.L. and Nagel, R.N. (1993). "Management, Technology and Agility: The Emergence of a New Era in Manufacturing," *International Journal of Technology Management* 8: pp. 18–38.

Gringeri, C.E. (1994). "Assembling 'Genuine GM Parts': Rural Homeworkers and Economic Development," *Economic Development Quarterly* 8: pp. 147–57.

Gros-Pietro, G.M. and Rolfo, S. (1989). "Flexible Automation and Firm Size: Some Empirical Evidence on the Italian Case," *Technovation* 9: pp. 493–503.

Hamel, G. and Prahalad, C.K. (1994). *Competing for the Future.* Boston: Harvard Business School Press.

Hansen, N. (1991). "Factories in Danish Fields: How High-Wage, Flexible Production Has Succeeded in Peripheral Jutland," *International Regional Science Review* 14: pp. 109–32.

Harrison, B. (1994). *Lean and Mean: The Changing Landscape of Corporate Power in the Age of Flexibility*. New York: Basic Books.

Harvey, D. (1989). *The Condition of Postmodernity*. London: Basil Blackwell.

Hayes, R.H. and Pisano, G.P. (1994). "Beyond World-Class: The New Manufacturing Strategy," *Harvard Business Review* 72 (1): pp. 77–86.

Hayes, R.H. and Wheelwright, S.C. (1984). *Restoring our Competitive Edge: Competing through Manufacturing*. New York: John Wiley.

Hiraoka, L.S. (1989). "Japanese Automobile Manufacturing in an American Setting," *Technological Forecasting and Social Change* 35: pp. 29–49

Hitchens, D.M.W.N.; Birnie, J.E.; and Wagner, K. (1992). "Competitiveness and Regional Development: The Case of Northern Ireland," *Regional Studies* 26: pp. 106–114.

Hoffman, K. and Kaplinsky, R. (1988). *Driving Force: The Global Restructuring of Technology. Labor. and Investment in the Automobile and Components Industries*. Boulder, CO: Westview Press.

Hudson, R. (1989) "Labour-Market Changes and New Forms of Work in Old Industrial Regions: Maybe Flexibility for Some but Not Flexible Accumulation," *Environment and Planning D: Society and Space* 7: pp. 5–30.

International Institute for Labour Studies [IILS] (1993). *Lean Production and Beyond: Labour Aspects of a New Production Concept*. Geneva: IILS.

Jaikumar, R. (1986). "Post-Industrial Manufacturing," *Harvard Business Review* 64 (6): pp. 69–76.

Johannisson, B. (1990). "University Training for Entrepreneurship: Swedish Approaches," *Entrepreneurship and Regional Development* 3: pp. 67–82.

Kenney, M. and Florida, R. (1988). "Beyond Mass Production: Production and the Labor Process in Japan," *Politics and Society* 16: pp. 121–58.

Kline, S.J. and Rosenberg, N. (1986). "An Overview of Innovation." In R. Landau and N. Rosenberg (Eds.), *The Positive Sum Strategy*. Washington: National Academy Press, pp. 275–305.

Kodama, F. (1992). "Technology Fusion and the New R&D," *Harvard Business Review* 70 (4): pp. 70–8.

Kristensen, P.H. (1989). "Denmark: An Experimental Laboratory for New Industrial Models," *Entrepreneurship and Regional Development* 1: pp. 245–55.

Kumpe, T. and Bolwijn, P.T. (1994). "Toward the Innovative Firm—Challenge for R&D Management," *Research-Technology Management* 37 (1): pp. 38–44.

Larsson, S. (1993). "New Dimensions in Organizing Industrial Networks," *International Journal of Technology Management* 8: pp. 39–58.

Lenz, J.E. (1989). *Flexible Manufacturing Benefits from the Low-Inventory Factory*. New York: Marcel Decker.

Ligus, R. (1990). "The Agile Enterprise," *Manufacturing Engineering* 105 (3): pp. 120.

Lloyd, P.E. (1989). "Research and Policy Review 28. Fragmenting Markets and the Dynamic Restructuring of Production: Issues for Spatial Policy," *Environment and Planning A* 21: pp. 429–44.

Lundvall, B.A. (1988). "Innovation as an Interactive Process: From User-Producer Interaction to the National System of Innovation." In G. Dosi, C. Freeman, R. Nelson, G. Silverberg, and L. Soete (Eds.), *Technical Change and Economic Theory*. London: Pinter, pp. 345–69.

MacPherson, A. (1988). "New Product Development Among Small Toronto Manufacturers: Empirical Evidence on the Role of Technical Service Linkages," *Economic Geography* 64: pp. 62–75.

_____ . (1991). "Interfirm Information Linkages in an Economically Disadvantaged Region: An Empirical Perspective from Metropolitan Buffalo," *Environment and Planning A* 23: pp. 591–605

_____ . (1992) "Innovation, External Technical Linkages and Small Firm Commercial Performance: An Empirical Analysis from Western New York," *Entrepreneurship and Regional Development* 4: pp. 165–82.

Maillat, D. (1991). "The Innovation Process and the Role of the Milieu." In E.M. Bergman, G. Maier; and F. Todtling (Eds.), *Regions Reconsidered.* London: Mansell, pp. 103–17.

Malecki, E.J. (1991). *Technology and Economic Development: The Dynamics of Local, Regional and National Change*. London: Longman, New York: John Wiley.

Malecki, E.J. and Todtling, F. (1994). "The New Flexible Economy: Shaping Regional and Local Institutions for Global Competition." In C.S. Bertuglia, M.M. Fischer, and G. Preto (Eds.), *Technological Innovation: Economic Development and Space.* Berlin: Springer, in press.

Malecki, E.J. and Tootle, D.M. (forthcoming). "The Role of Networks in Small Firm Competitiveness," *International Journal of Technology Management*, in press.

Malecki, E.J. and Veldhoen, M.E. (1993). "Network Activities, Information and Competitiveness in Small Firms," *Geografiska Annaler* 75B: pp. 131–47.

Malerba, F. (1992). "Learning by Firms and Incremental Technical Change," *Economic Journal* 102: pp. 845–869.

Manufacturing Engineering (1993). "In Search of Agile Manufacturing," 110 (2): pp. 24–6.

Martinelli, F. and Schoenberger, E. (1991) "Oligopoly Is Alive and Well: Notes for a Broader Discussion of Flexible Accumulation." In G. Benko and M. Dunford (Eds.), *Industrial Change and Regional Development.* London: Belhaven, pp. 117–33.

McArthur, R. (1989). "Locality and Small Firms: Some Reflections from the Franco-British Project, 'Industrial Systems, Technical Change and Locality,'" *Environment and Planning D: Society and Space* 7: pp. 197–210.

McCutcheon, D.M.; Raturi, A.S.; and Meredith, J.R. (1994). "The Customization-Responsiveness Squeeze," *Sloan Management Review* 35 (2): pp. 89–99.

Miller, H.M.; Brown, E.E.; and Centner, T.J. (1987). "Southern Appalachian Handicrafts Industry: Implications for Regional Economic Development," *Review of Regional Studies* 17 (3): pp. 50–8.

Møller, T. (1991). "Technology and Growth—Measures and Concepts: A Case Study of Denmark," *Technovation* 11: pp. 475–81.

O'Farrell, P.N. and Hitchens, D.N.M.W. (1988). "The Relative Competitiveness and Performance of Small Manufacturing Firms in Scotland and the Mid-West of Ireland: An Analysis of Matched Pairs," *Regional Studies* 22: pp. 399–416.

_____ . (1989). "The Competitiveness and Performance of Small Manufacturing Firms: An Analysis of Matched Pairs in Scotland and England," *Environment and Planning A* 21: pp. 1241–63.

Pine, B.J. (1993). *Mass Customization: The New Frontier of Business Competition*. Boston: Harvard Business School Press.

Piore, M. and Sabel, C. (1984). *The Second Industrial Divide*. New York: Basic Books.

Prahalad, C.K. and Hamel, G. (1990). "The Core Competence of the Corporation," *Harvard Business Review* 68 (3): pp. 79–91.

Pyke, F. (1992). *Industrial Development Through Small-Firm Cooperation: Theory and Practice*. Geneva: International Labor Office.

Pyke, F.; Becattini, G.; and Sengenberger, W. (Eds.) (1990). *Industrial Districts and Inter-Firm Cooperation in Italy*. Geneva: International Institute for Labour Studies.

Quinn, J.B. and Hilmer, F.G. (1994). "Strategic Outsourcing," *Sloan Management Review* 35 (4): pp. 43–55.

Rothwell, R. (1991). "External Networking and Innovation in Small and Medium-Sized Manufacturing Firms in Europe," *Technovation* 11: pp. 93–112.

Sayer, A. (1989). "Postfordism in Question," *International Journal of Urban and Regional Research* 13: pp. 666–95.

Schmitz, H. (1992). "Industrial Districts: Model and Reality in Baden Wurttemberg, Germany." In F. Pyke and W. Sengenberger (Eds.), *Industrial Districts and Local Economic Regeneration*. Geneva: International Institute for Labour Studies, pp. 87–121.

Sciberras, E. (1986). "Technical Innovation and International Competitiveness in the Television Industry." In E. Rhodes and D. Wield (Eds.), *Implementing New Technologies*. Oxford: Basil Blackwell, pp. 177–90.

Scott, A.J. (1988). *New Industrial Spaces*. London: Pion.

Shaiken, H. (1984). *Work Transformed: Automation and Labor in the Computer Age*. New York: Holt, Rinehart and Winston.

Shaw, B. (1991). "Developing Technological Innovations within Networks," *Entrepreneurship and Regional Development* 3: pp. 111–28.

Sheridan, J.H. (1993). "Agile Manufacturing: Stepping Beyond Lean Production," *Industry Week* 242 (8) April 19: pp. 30–46.

Smilor, R W. and Feeser, H.R. (1991). "Chaos and the Entrepreneurial Process: Patterns and Policy Implications for Technology Entrepreneurship," *Journal of Business Venturing* 6: pp. 160–70.

Starr, M.K. and Biloski, A.J. (1985). "The Decision to Adopt New Technology—Effects on Organizational Size." In E. Rhodes and D. Wield (Eds.), *Implementing New Technologies*. London: Basil Blackwell, pp. 303–15

Stevens, B. and Andrieu, M. (1991). "Trade, Investment and Technology in a Changing International Environment." In OECD, *Trade, Investment and Technology in the 1990s*. Paris: Organisation for Economic Co-operation and Development, pp. 113–25.

Stohr, W.B. (1986). "Regional Innovation Complexes," *Papers of the Regional Science Association* 59: pp. 29–44.

Storper, M. (1993). "Regional 'Worlds' of Production: Learning and Innovation in the Technology Districts of France, Italy and the USA," *Regional Studies* 27: pp. 433–55.

Storper, M. and Scott, A.J. (Eds.) (1992). *Pathways to Industrialization and Regional Development*. London: Routledge.

Storper, M. and Walker, R. (1989). *The Capitalist Imperative: Territory. Technology, and Industrial Growth*. Oxford: Basil Blackwell.

Teece, D.J. (1980). "Economies of Scope and the Scope of the Enterprise," *Journal of Economic Behavior and Organization* 1: pp. 223–47.

_____. (1986). "Profiting from Technological Innovation: Implications for Integration, Collaboration, Licensing and Public Policy," *Research Policy* 15: pp. 285–305.

Thorelli, H.B. (1986). "Networks: Between Markets and Hierarchies," *Strategic Management Journal* 7: pp. 37–51.

Upton, D.M. (1994). "The Management of Manufacturing Flexibility," *California Management Review* 36 (2): pp. 72–89.

Utterback, J.M. (1994). *Mastering the Dynamics of Innovation*. Boston: Harvard Business School Press.

Walton, R.E. and Susman, G.I. (1987). "People Policies for the New Machines," *Harvard Business Review* 65 (2): pp. 98–106.

Warner, T.N. (1987). "Information Technology as a Competitive Burden," *Sloan Management Review* 29 (1): pp. 55–61.

Womack J.P.; Jones, D.T.; and Roos, D. (1990). *The Machine That Changed the World*. New York: Rawson Associates.

Young, R.C. and Francis, J.D. (1989). "Who Helps New Manufacturing Firms Get Started?" *Rural Development Perspectives* 6 (1): pp. 21–5.

Youssef, M.A. (1992). "Agile Manufacturing: A Necessary Condition for Competing in Global Markets," *Industrial Engineering* 24 (12):1 pp. 8–20.

Zeitlin, J. (1992). "Industrial Districts and Local Economic Regeneration: Overview and Comment." In F. Pyke and W. Sengenberger (Eds.), *Industrial Districts and Local Economic Regeneration*. Geneva: International Institute for Labour Studies, pp. 279–94.

3

LABOR AND CAPITALIST ACCUMULATION IN THE LATE TWENTIETH CENTURY: WHAT AND WHO IS FLEXIBLE?

Martin Kenney
Department of Applied Behavioral Sciences, University of California at Davis, Davis, CA 95616 USA
and
James Curry
El Colegio de la Frontera Norte, Tijuana, Baja California, MEXICO, (P.O. Box L, Chula Vista, CA 91912 USA)

INTRODUCTION

A recent *Wall Street Journal* article reported the new UAW assembly line workers that Chrysler and Ford are currently hiring are far better-educated than their current work force. The UAW is opposed to this effort to hire better-educated workers. Ex-UAW president Fraser was quoted as saying "Let's face it, it doesn't take a rocket scientist to work on the assembly line (Templin, 1994)." During the last decade of the twentieth century there have been intense changes underway in the global political economy and these changes are profoundly impacting how we think about labor and labor relations. The proximate cause for this rethinking is the remarkable success of Japanese capitalism. Trying to understand the changes labor is experiencing is, in effect, to try to explain the changes in capitalism itself. As Sayer and Walker (1992) aptly point out, there is a reworking and a new, more powerful integration of the global division of labor. This is reflected in and is an aspect of a far deeper and more profound globalization of the wage labor relationship. Importantly, though capitalism has always accumulated on a world scale, in the current period it is penetrating new regions with the intent of integrating the inhabitants into the global marketplace as producers. As importantly, like a virus, it is constantly mutating its surface coating (i.e., labor relations, interfirm relations, the role of managers etc.), while reproducing its essential genetic code. Thus, its tremendous flexibility.

In recent years there has been an outpouring of work attempting to create models that capture the tenets of an emerging stable political economic system. There are number of different streams of explanation: The first and, perhaps, most influential stream is associated with the French Regulation School which borrowed Antonio Gramsci's term "Fordism" to describe the postwar global stability.[1] Recently, there has been a proliferation of terms such as post-Fordism, hyper-Fordism, peripheral Fordism, Toyotism, Ohnoism (after the inventor of Toyota's kanban system), and neo-Fordism (Lipietz, 1987; Coriat, 1991; Dohse, Jurgens, and Malsch, 1986). A second and related stream branching from regulation theory revolves around some permutation of the word "flexibility." The most celebrated proponents of this are Piore and Sabel (1984) who coined the term "flexible specialization." Harvey (1985, 1989) nearly contemporaneously argued that capitalism was developing a system of "flexible accumulation." As Sayer (1989), initially, and later Curry (1993) pointed out, the word "flexible" has been used vaguely and refers to a wide range of conditions. This has made it difficult to concretize. Most human institutions have and need elements of "flexibility" and "rigidity." In an even more recent article, Upton (1994) argues that there are numerous dimensions of flexibility, and identifying the type of flexibility which suits a particular firm is vital for economic survival. The conclusion to be drawn is that if flexibility or flexible accumulation are to have meaning, then they must be far more rigorously specified.

Yet another intellectual stream aiming to give coherence to the changes in the institutions and arrangements of the global political economy is the one associated with Perez and Freeman (1985) which seeks to explain the changes in terms of a newly emerging techno-economic paradigm. The merit of this argument is that it explicitly integrates the role of technology in the changing global economy. As with the other perspectives, the techno-economic paradigm school is vague regarding the institutions and concrete mechanisms which are driving the new paradigm, resting the bulk of the causality on the emerging information technologies. On the other hand, Perez and Freeman (1985) were not trapped into discussion on the nature of flexibility or the francophone desire to coin terms for a new successor regime for Fordism.[2]

To comprehend the complexity of the current changes in labor relations it is useful to use the less grandiose but more usable concept of commodity chains (Hopkins and Wallerstein, 1986; Gereffi, 1992). This is necessary because labor relations are predicated upon production, but, increasingly, production is part of a global division of labor (Sayer and Walker, 1992). Commodity chain analysis directs attention to the different segments or modules of a production process, this encourages examination of how different production sites often undertake only one segment of the entire commodity chain or production process. This is because different steps in the production of a single product may require or allow the use of quite different types of labor, labor organization, technology, etc. Often, these different steps need not be collocated. Thus, at each site it is possible to maximize or minimize particular variables such as labor cost, skills, and infrastructure, among others. Moreover, these divisions of labor are not static, rather they are tentative and fleeting. Each node is open to revision through growth, decline, or termination.

To explore the implications of these changes in the global political economy this chapter is divided into four sections. The first and second sections in stylized terms outline the contours of the work process for unionized workers in the U.S. and Japanese firms, respectively. The second section also explores some of theoretical frameworks that have been proffered to explain Japanese success and then continues by examining the dimensions of the transfer of some of the central features of the Japanese system to the U.S. This section sets the stage for examining the current rethinking of the importance of production workers for the successful firm. The core of these current developments is an attempt to more fully integrate the worker into the firm and its activity. The third section considers the type of worker necessary in factories using the new management techniques. The fourth section turns to a different reality, that of the rise of sweatshops in even advanced developed countries. In these establishments the worker is merely a pair of hands doing highly repetitive work. The aim is to capture the other face of the "new" capitalism. The conclusion grapples with the complexity of understanding the evolving nature and role of labor in the current global economy.

TRADITIONAL U.S. LABOR RELATIONS

The traditional Fordist U.S. factory was based on the relatively unskilled workers manning a station on an semi-automated assembly line. The workers were functionally specialized and required to complete one or two highly routinized physical tasks in a discrete time period. Beyond these physical tasks the worker had few other responsibilities. The actual production process was encased within the larger framework of a conflictual relationship between management and labor. This ongoing conflict had resulted in the building of an intricate and even arcane structure of job classifications and rules which both protected the worker and constrained both management and labor's ability to change the production process (see, for example, Florida and Kenney, 1992).

Sayer (1986) termed the Fordist production process as the "just-in-case" system. Throughout the factory and in the warehouse there was inventory in case something went wrong. Workers were responsible for quantity and not quality. Quality inspectors were a separate group of white-collar employees whose sole function was to measure quality. The result was enormous amounts of rework with little improvement, because quality controllers were so busy inspecting that they took no measures to improve quality.

By the 1970s, the crisis of the quintessential American Fordist industry, automobile production, was fully manifest. As Aronowitz's (1973) study of the GM plant in Cleveland, Ohio showed, labor discipline had broken down. Factories were characterized by substance abuse, physical conflict, and rampant absenteeism. The workplace was dirty, unsafe, and unsanitary. Management was abusive and capricious. Labor and the unions protected and even encouraged incompetent and recalcitrant workers. It was surprising that any products could be manufactured in U.S. factories.[3]

The Fordist factory and the form of labor organization it represented was already in crisis when, after initial setbacks, the Japanese auto industry began to make inroads into the U.S. market. With superior quality and productivity (and aided by a weak American corporate response to emerging gasoline shortagaes), Japanese firms quickly gained market share and by the early 1980s had driven Chrysler to the verge of bankruptcy. U.S. firms responded by mass lay-offs, factory closures, and by pressing for protection. None of these measures proved effective, because the Japanese industry was using a more advanced production system which enabled them more effectively to utilize their human resources (Womack, Jones, and Roos, 1990; Kenney and Florida, 1993).

THE JAPANESE PRODUCTION SYSTEM

The primary features of the Japanese production system were developed in the aftermath of Japan's defeat in World War II (for further discussion see Kenney and Florida, 1988; Gordon, 1985; Kato and Steven, 1993).[4] The immediate postwar period saw a rapid increase in the power of radical unions and a drastic reordering of the role of workers in the firm. Simultaneously, there was a decisive weakening of the role of stockholders (owners) who were replaced by professional managers. By 1950 the radical unions had been decisively defeated and more compliant enterprise unions had become dominant in many, but not all, firms.

The major features of the Japanese production system consist of the following: enterprise unions, wages determined primarily by seniority, and long-term employment.[5] Each of these organizational components provided elements of "flexibility" and "rigidity." For example, the long-term employment feature creates a fundamental rigidity because in a downturn the company is unable to terminate excess workers. Long-term employees are a fixed cost. According to recent press accounts concerning the current recession, this has resulted in Japanese firms carrying over one million redundant workers. Conversely, long-term employment provides some important flexibilities. The first is that, even though Japanese workers are guaranteed employment, they are not guaranteed a specific job. Therefore, they can be shifted from job to job within a plant or even moved to other plants and, in a worst case scenario, transferred to affiliate firms. Moreover, this constant pressure to keep workers employed places tremendous pressure on management to find new businesses in which to utilize their workers.

The long-term employment system also means that a company is guaranteed that its investment in training employees can be amortized over the employee's career (see Koike, 1988; Aoki, 1988). In more open labor markets the newly trained employee might easily choose to leave the firm that trained him/her for another firm. Once again, this has a rigid aspect and flexible aspect. The newly trained worker should be more capable. However, the company's investment in training is only economically viable because of labor market rigidity.

If enterprise unionism, seniority wage, and long-term employment form the macrocontours of the Japanese system, then it is teams that form the core of the production process. As Fruin (1992) points out, the focus of the Japanese factory is the team. More generally, Fruin argues that all the activities of the Japanese firm are organized on the basis of teams. Team organization allows many flexible staffing patterns such as rotating team members from job to job thereby spreading the most onerous jobs and cross-training workers. On the other hand, a team-based system can lead to "group think" which suppresses some individual capabilities and can inhibit innovation and thereby lead to rigidity. The point being that simple polarities between flexibility and rigidity cannot encompass the Japanese system of organizing the work process and managing labor relations.

The team, not the individual, is the fundamental organizational principle of the Japanese factory. The team is responsible for a set of tasks and its members are expected to work together to accomplish the tasks. The team must not only apportion the tasks but also must participate in continuously improving the production process through kaizen activity. Adler and Cole (1993) point out that at the NUMMI automobile assembly facility in Fremont, California the team is actively involved in developing the work standards and changing them through improvements. In this sense, the workers are not only learning, but they are also creating new knowledge about the production process. Thus, whereas the traditional assembly-line was static, the Japanese assembly line constantly evolves due to workers' own improvement of the production process and, of course, input by engineers.

The team organization also provides a way for management to put increased pressure, i.e., increased speed, higher quality standards, decreased quantity of inputs, etc., on the production system. This is a subtle point, but essentially Japanese managers discovered that workers functioning as a team were able to accomplish more than workers operating as individuals. If, for example, a new task was added to the bundle for which the team was responsible, often the team could rebalance the tasks of the group to incorporate the new task. Or, conversely, management could accelerate the production line and the team itself would innovate solutions (kaizen) to cope with the greater speed. The individual worker is also responsible for ensuring that no defective parts pass his station, a task that all Japanese factory workers have. This constant addition of new tasks and thereby stressing the system has led Parker and Slaughter (1988) to term the Japanese system "management-by-stress." Indeed, Parker and Slaughter have captured one facet of the system. However, if the system merely made workers toil harder, it would not be the powerful system that it has proven to be. The key issue is that the Japanese system constantly pressures workers (through the medium of the team) to work, innovate and improve the process (or what Kenney and Florida, 1993, have termed "innovation-mediated production").

The team also provides other advantages to management. Because the team is responsible for a certain set of activities, if any team member is missing, the weight of completing the required work falls on the other members of the team (especially the team leader who must substitute for the missing worker). When the team leader works, the other team members must also work harder by absorbing the leader's tasks. This arrangement places enormous pressure upon workers to come to work every day, because their team members will bear the immediate brunt of replacing the worker. Thus, routine discipline on tardiness and absenteeism is enforced without the overt role of management.

However, what Parker and Slaughter ignore is that the system can also improve the worker's situation.[6] There are macro- and microdimensions to this improvement. The macrodimensional improvement is that the workers have a limited, but nevertheless real empowerment in the work process. In a sense, capital has ceded a certain measure of overt power to gain workers' more total cooperation. Conversely, through organizational arrangements such as the JIT system, workers have far more potential than ever before to disrupt the work process. The microdimensional improvement is a result of the fact that worker improvement occurs not only in higher productivity, but also includes issues such as health and safety. Here, this limited empowerment can provide substantial benefits to the workers (for discussion of this see Kenney and Florida, 1993).

Upon this base Japanese firms began their small group activities such as quality control circles (QCC) in the early 1960s (see Cole, 1989; Kyotani, 1994). There are two interpretations of QCC. One is that the quality control movement was oriented by the technical benefits it could yield. But there is a second more radical explanation, which is that it was part of a massive corporate effort to remove power from the factory floor unions by enrolling workers in the management-guided QCCs. This issue is now the topic of some debate in Japanese social science circles. However, regardless of the initial intentions by management, there is no doubt that QCC and other small group activity movements have been successful in improving efficiency and quality, while also performing the function of providing workers with some feeling of empowerment. Moreover, in the process of participating in these small group activities workers both exercise their intellectual capabilities and learn more about their role in the activities of the entire plant. And, in this sense, the worker and the plant become more flexible.

It is not the "flexibility" of the workers that is the key to Japanese success, rather it is the way in which Japanese industry has been able to orient and use their mental capabilities to improve production. This has prompted scholars to think about the factory, not only in terms of physical production but also as a learning system or as Fruin (1992) terms Toshiba's Yanagicho factory, a "learning" works. In other venues, Kenney and Florida (1993) reconceptualize the factory as a "laboratory." This is because QCC activity is a special form of learning, which actually encourages experimentation with different ways of accomplishing activities And like a laboratory experiment standardized work is, in actuality, a protocol for reproducibility. It is the reproducibility that allows verification and improvement.[7] In a laboratory there is a premium on exactness and thought.

If the previous characterization of the work process is correct, then a quite different type of labor relations is required. The emphasis on learning or knowledge creation is one method of thinking about the changing nature of the capitalist worker. Koji Kobayashi, the chairman of NEC, has termed this new knowledge worker (referring especially to engineers) as "V-shaped" in their knowledge base, that is, the worker is initially trained as a specialist with a very deep knowledge in a narrow area. But, after that they should spread their knowledge more widely, thus the V-shape, to their capabilities. Roobeek (1987) writes that workers in large companies will receive a "reliability" wage. Though this is a rather passive formulation, it captures the dependence that large firms have on their production employees.

One important result of Japanese success in the global economy has been to demonstrate the importance of labor in achieving successful production. Moreover, Japanese success has shown that Braverman's argument, which did actually capture the belief of U.S. management that contemporary capitalism was based on a fundamental disassociation of conception and execution, was inherently a flawed conceptualization. The erroneousness of this belief in this dichotomy also has been proven in intercapitalist competition. It is this competitive failure which is prompting leading U.S. firms to revamp their strategies regarding labor.

THE NEW KNOWLEDGE WORKER IN AMERICA

Japanese success in steel, automobiles, consumer electronics, and then electronics was initially explained on the basis of lower wages, weak labor unions, or a culture of compliance. These positions were tenable until Japanese firms established factories overseas and successfully exported a significant portion of their labor-management relations and production system to the U.S. (Kenney and Florida, 1993; Abo, 1994). The most important Japanese investments were in the quintessential Fordist industry, automobiles. And, it was the GM-Toyota joint venture, New United Motors Manufacturing Inc. (NUMMI), in Fremont, California, which opened in 1984 that forced U.S. management to rethink the fundamental premises of labor-management relations (Krafcik, 1986; Brown and Reich, 1989).

The GM Fremont plant had been one of the worst in the entire GM system in terms of quality, labor relations, and productivity. When the Fremont plant closed there were more than 1,000 unsettled grievances still pending (Krafcik, 1986). Not only were labor-management relations abysmal, but absenteeism (at times over 20%), tardiness, sabotage, alcohol and drug usage on the job were rampant (NUMMI, 1986). The plant was stereotypical of the collapse of the Fordist system.

NUMMI was such a crucial experiment because Toyota reopened and accepted management responsibility for the Fremont plant which GM had closed. Moreover, NUMMI agreed to be unionized by the UAW and rehired almost all of the workers previously laid off who wished to return. However, the new labor contract was entirely different from previous UAW contracts on a number of dimensions. For example, NUMMI would have only four job classifications (at the time its sister GM plant in Framingham, Massachusetts had 183) and all hourly workers were organized into teams (Krafcik, 1986). From the beginning, UAW members were actively involved in planning their jobs and standardizing the work (Adler, 1993). The union contract specified that workers would be involved in kaizen or continuous improvement activities aimed at making the company more efficient (UAW, 1991). The upshot was a radical transformation of the old Fremont plant. NUMMI became one of the highest quality, most productive plants in North America. Drug and alcohol abuse were also drastically reduced and workers were far more satisfied with their jobs.

If the saga of NUMMI were an isolated occurrence, it would remain an exciting tale of little further interest. But, as Kenney and Florida (1993) argue there are many cases in which the Japanese companies have been successful in transferring important elements of their labor and production system to the U.S. (for slightly less sanguine position, see Abo, 1994).

The success of Japanese firms in transferring their labor-management relations to the U.S. has exerted a powerful demonstration effect. An increasing number of U.S. firms such as Hewlett-Packard, Motorola, Xerox, Ford, and Chrysler are experimenting with organizational techniques and concepts such as teamwork, continuous improvement, and new relations between management and labor. And it seems likely that the adoption of these new industrial relations will be necessary for the survival of these firms in the Information Age (for further discussion, see Zuboff, 1988).

There seems little doubt that a new knowledge worker is emerging on the factory floor in a number of firms around the world. The fundamental question is whether this category of worker will become dominant or if they are only an aristocracy of the workforce. Put somewhat differently, is the knowledge worker the dominant emerging pattern or are other patterns also emerging?

THE NON-THINK WORKERS

If there is one group of workers that receive significant training, there is another group that operate in a far more Taylorist environment. Many are, in large measure, merely casuals that can be disposed of at no loss to the firm. These consist of what Berger and Piore (1980) termed the periphery of the labor market. They provide the firm with the greatest possible flexibility as there is no commitment to nor investment in these workers.

Previously in the U.S. and Europe, workers in large firms had stable jobs protected by unions or company norms. But the development of improved communication and transportation plus advances in production techniques now permit greater degrees of freedom in the location of production facilities. Most importantly, these developments have allowed firms to partition the assembly line into modules. Each of these modules could be located in environments most suitable to overall cost and risk minimization. As an example, color television (CTV) production has been divided into modules that require very different mixes of capital, labor type, and labor quality. So, television assembly and packing requires low capital investment and is labor-intensive. Thus, most television assembly can be and is located in low-wage environments and can be quickly relocated if wages increase. Conversely, printed circuit board assembly (PCB), which was formerly an extremely labor-intensive process located in low-wage environments, has become increasingly automated and capital-intensive. Thus, though wages are still a factor, automation can permit the PCB assembly process to be relocated from developing to developed countries. An excellent example of this is Solectron, the largest contract manufacturer of PCBs, which undertakes much of its PCB production in relatively high labor cost Silicon Valley.[8]

CTV tube production is extremely capital-intensive and requires trained technicians and engineers. Therefore, a sophisticated infrastructure and stable social formation capable of producing high-quality engineers and technicians is necessary. Moreover, because of the large capital investment, the tube facility is not easily abandoned, thus the firm has much less flexibility in relocating. Until recently, tube production was largely confined to the developed countries. In the commodity chain for manufacturing a CTV these three important links in the chain demand different local labor market characteristics.

A type of production work that currently is not treated as "think" work is that done in the Mexican maquiladoras. Turnover and absenteeism are high. For example, in 1991 in Japanese maquiladoras turnover rates ranged from 4–15% per month, the mode being 10% per month (Kenney and Florida, 1994). In 1988 the U.S.-owned maquilas experienced turnover rates of between 10–20% per month (Sklair, 1989). One maquiladora interviewed in 1993 reported that its initial turnover rate was 30–40% per month, but later settled to 10% per month (Kenney et al., 1994).

Ordinarily, such high rates of turnover are considered a severe problem. Yet the president of a Japanese supplier maquila reported that: "From a certain level [of employee] we listen. . . . Below that, we consider the turn-over zone. So, we do not even listen to them. . . . Listening to [those in] the turnover zone just confuses the operation." Clearly, this Japanese company had invested little in most of its workers. They were only hands. Conversely, perhaps 15% of the workers could contribute on the basis of their knowledge.[9] Actually, this labor relations philosophy is not entirely different from that of Japan, as Japanese companies also hire part-time and temporary workers who are expected to be significantly less integrated into the firm's activities.

In yet another example of the contradictory nature of the current global changes in labor relations, Diaz (1993) shows in a study of Chile (a country that has recently been touted as one of the success stories of the 1980s) that the "flexibility" now extant in the Chilean economy is better characterized as the informalization of much of the working population. Thus, for the vast majority of Chilean workers, the onset of flexibility is hardly the panacea that Piore and Sabel (1984) theorized.

This casualization or informalization with its concomitant flexibility has developed not only in the Third World, but also in the developed countries with the continuing increase in low-paid service workers, temporary help agencies and sweatshops. These low-paid workers provide value for accumulation. However, there are serious questions as to whether workers operating in fields with low wages are, for example, capable of paying the taxes necessary to reproduce an advanced developed country's infrastructure.

And yet, there is also evidence that it is not impossible to move up the production ladder as Taiwan, South Korea, and Malaysia have shown. Moreover, in these countries there has been an increasing development of skills and a stabilization in the workforce. Whether this is due to flexible accumulation or rather to an integration of these economies into the Japanese production system is difficult to say. However, these countries with the exception of Korea, have been the recipients of large-scale Japanese investment and all have fashioned a number of institutions that resemble those of Japan.

THE CONCEPT OF FLEXIBLE ACCUMULATION RECONSIDERED

The global economy is in flux and there is increased mobility on the part of capital but the mobility of capital is not absolute. In some cases it is high (see, for example, Gereffi, 1992), while in other cases it is quite low (Harvey, 1989). For example, in the U.S. Midwest Japanese heavy industries have invested more than $15 billion in plant and equipment (Kenney and Florida, 1993). These investments, which have contributed greatly to the industrial revival of the Midwest, are not highly mobile.

This essay has briefly highlighted some of the major issues concerning labor relations in the contemporary globalizing economy. Though these changes may not conform with the flexible accumulation paradigm proposed by Harvey (1989), there is convincing evidence of an epoch-making change underway. What is most interesting from the labor relations standpoint is how there can be two radically different tendencies underway simultaneously. In one tendency, worker training and involvement are considered critical for quality, efficiency and effectiveness. In the other tendency, workers are considered to be little more than hands.

The coexistence of two apparently divergent poles of labor organization raises the question of how they are related, (if indeed they are related). Why does highly routinized sweated labor continue to exist within the context of an opposite tendency which seems to be requiring more worker participation and creativity?

The garment industry holds some clues. Garments continue to be produced in advanced developed countries because of two dominant conditions and one minor condition: The first condition, as Barff and Austen (1993) demonstrate in the case of athletic shoes, is a favorable tariff protection regime. The second, and far more common, condition is the existence of a large pool of cheap legal and illegal immigrant labor, such as is found in the Los Angeles area. This has been the key factor in the development of sweatshop garment manufacturing in the U.S. That the technology of manufacture is stable and non-complex contributes to the bifurcation of the garment industry into a low-knowledge manufacturing component and a high-knowledge design and marketing component. The third rather less powerful condition is that some consumers are willing to pay a premium price for an extremely high quality product. In such cases, skilled labor can continue to survive in the developed countries, and provide a developed country standard of living, even in the garment industry.

More generally there are three reasons for the continuance of low-wage, low-skill jobs in the advanced developed countries. First, some processes are extremely hard to automate (for instance, garment and some assembly operations) because the physical form of the product and the methods of manufacture are not amenable to machine production. In other words, due to either technical factors such as complex, not entirely predictable inputs, or economic factors such as the price of the commodity or the value composition of the commodity, it is not cost-effective to utilize automation or advanced production methodologies. Second, for some processes automation is undesirable. In these situations the non-automated product is more marketable than the automated product. For example, many people go to a restaurant for, at least in part, the experience of being served by another person. It is no doubt possible to automate the process much further than it is currently, say, through the installation of an conveyor belt that directly links the table to the kitchen. Such a system might work well for busy workday lunch customers but is unlikely to be successful among people who are dining out for the entertainment experience. The third instance consists of processes which are inherently knowledge-intensive and cannot be automated or routinized, or done so only in a limited fashion (e.g., primarily creative activities such as software writing, fashion design, marketing, management and coordination, production of entertainment goods such as books, movies, music, etc.).[10]

It is not enough to say the new capitalist accumulation is flexible; capitalist accumulation has always been flexible (and anarchic). The combination of skilling, re-skilling, and increasing flexibility is an expression of the real movement of capitalism. This reality is often obscured by the preoccupation with broad descriptive terminology such as Fordism or flexible specialization.

The new global configuration combines different types of production methodologies. While some might refer to this as flexibility, it might better be considered as a type of differentiation. This diffentiation however is not fixed, but rather is constantly in flux being redrawn, rethought, and rearranged. The changes in advanced capitalist production location which some attribute to postmodern fragmentation are more effectively explained as the result of the continuous reexamination of the options presented to managers. Thus, these changes are the results of conscious strategies by specific firms responding to a rapidly changing landscape of conditions and opportunities. This is also occurring at the product level not only by inventing entirely new products, but by giving mass produced products a veneer of variety (Pine, 1993, has termed this "mass customization").

Some new products are coincident with the development of new product technologies or new production technologies (or a combination of the two). But differentiation of products is not necessarily dependent on differentiation of production or production methodologies. Styles of clothing change far more rapidly than do the production techniques or materials used to produce them. For many "low technology" products, such as garments, much of the value of the product is "added" at the design stage. Complex production technologies such as computer-aided design may be used for design, but they are often too expensive relative to the conditions of the market at the production stage. Sweatshop subcontracting makes more economic sense. Once a "hot" design is developed, the problem is to reduce it to production and get it to market as quickly as possible. The main technological barrier is translating the language of design into the language of production. Advanced technologies have accelerated and improved communication, but the human component remains crucial. Thus, often advanced production technologies and methodologies are combined with more primitive ones in the same commodity chain. This combination means it is more difficult than ever to refer to a single dominant form of labor relations system emerging.

ENDNOTES

[1]For the early Regulation School, see Aglietta (1979). The later Regulation School is represented by Lipietz (1987) and Boyer (1990), who have tried to formalize regulation theory into a School. In the process, many adherents, especially Boyer, have sacrificed much of the theory's flexibility and innovativeness in a search for mathematical rigor. For a cogent explanation of regulation theory, see De Vroey (1984).

[2]French regulationists building upon Gramsci's fundamental insights have attempted to capture the essence of the newly emerging production system. Yet, they continue to focus upon the automobile. Their attempts to conceptualize the emerging system are variations on Fordism. This is interesting because, in their own schemata, the precursor to Fordism was Taylorism, which was based on the previous leading industry, steel. Nevertheless, their successor to Fordism is "Toyotism," obviously car-related or "Ohnoism" after Taiichi Ohno, the initial developer of the Toyota production system. A sociology of knowledge perspective may assist in understanding the reasons for this. Essentially, the milieu in which French regulationists think and operate is France, and Europe in general, both of which are relatively underdeveloped in electronics. This is unfortunate because it is the macroindustry of electronics which is rapidly replacing automobiles as the key industry of the future. For example, electronics and electrical parts already account for 20% of the value of a car.

[3]Hamper (1991) provides an excellent worker's level account of working on the GM assembly line.

[4]We recognize that many of the features that became generalized in the postwar period have antecedents in the prewar period. It is their generalization that we are referring to here.

[5]This entire discussion refers only to the regular workers in the large enterprises or between 25–35% of the workforce. Thus, temporary and part-time workers and those that work for smaller firms are omitted.

[6]For an excellent discussion of these and other issues see Bonazzi (1994).

[7]This closely parallels the argument by Adler (1993) about standardization being the basis upon which continuous improvement can work.

[8]Barff and Austen (1993) and Donaghu and Barff (1990) have documented a similar arrangement in athletic shoe production. A recent edited book by Gereffi and Korsenwieciwz (1994) contains a number of articles examining this increasingly complex division of labor.

[9]I would like to thank Charles Sabel for pointing this out.

[10]All of these bear resemblance to what Robert Reich (1991) termed symbol analysis.

REFERENCES

Abo, T. (Ed.). (1994). *Hybrid Factory: The Japanese Production System in the United States*. New York: Oxford University Press.

Adler, P. (1993). "Time-And-Motion Regained," *Harvard Business Review* 71, (1): pp. 97–109.

Adler, P. and Cole, R. (1993). "Designed for Learning: A Tale of Two Auto Plants," *Sloan Management Review* (Spring): pp. 85–94.

Aglietta, M. (1979). *A Theory of Capitalist Regulation*. London: New Left Books.

Aoki, M. (1988). *Information, Incentives and Bargaining in the Japanese Economy.* Cambridge: Cambridge University Press.

Aronowitz, S. (1973). *False Promises.* New York: McGraw-Hill.

Barff, R. and Austen, J. (1993). "'It's Gotta Be Da Shoes': Domestic Manufacturing, International Subcontracting, and the Production of Athletic Footwear," *Environment and Planning A* 25 (8): pp. 1103–14.

Berger, S. and Piore, M. (1980). *Dualism and Discontinuity in Industrial Societies.* Cambridge: Cambridge University Press.

Bonazzi, G. (1994). "Between Post-Fordism and the Japanese Model: Cognitive Processes in European and American Sociology," unpublished mimeo, Department of Social Sciences, University of Torino.

Boyer, R. (1990). *The Regulation School: A Critical Introduction.* New York: Columbia University Press.

Brown, C. and Reich, M. (1989). "When Does Union-Management Cooperation Work: A Look at NUMMI and GM-Van Nuys," *California Management Review* 31, (4): pp. 26–44.

Cole, R. (1989). *Strategies for Learning.* Berkeley: University of California Press.

Coriat, B. (1991). *Penser a l'Envers.* Paris: Christian Bourgois.

Curry, J. (1993). "The Flexibility Fetish: a Review Essay on Flexible Specialization," *Capital and Class* 51: pp. 49–76.

DeVroey, M. (1984). "A Regulation Approach Interpretation of the Contemporary Crisis," *Capital and Class* 23: pp. 45–66.

Diaz. A. (1993). "Industrial Governance in Latin America." Paper presented at the SSRC Workshop on Industiral Governance and Labor Flexibility in Comparative Perspective, September 17–19, New York, New York.

Dohse, K.; Jurgens, U.; and Malsch, T. (1986). "From Fordism to Toyotism? The Social Organization of the Labor Process in the Japanese Automobile Industry," *Politics and Society* 14, (2): pp. 45–66

Donaghu, M. and Barff, R.A. (1990). "Nike Just Did It: International Subcontracting and Flexibility in Athletic Footware Production," *Regional Studies* 24, (6): pp. 537–552.

Florida, R. and Kenney, M. (1992). "Restructuring in Place: Japanese Investment, Production Organization and the Restructuring of Steel," *Economic Geography* 68, (2): pp. 146–73.

Fruin, W.M. (1992). *The Japanese Enterprise System: Competitive Strategies and Cooperative Structures.* Oxford: Clarendon Press and New York: Oxford University Press.

Gereffi, G. (1992). "New Realities of Industrial Development in East Asia and Latin America." In R. Appelbaum and J. Henderson (Eds.), *States and Development in the Asian Pacific Rim,* Newbury Park, CA: Sage Publications, pp. 85–111.

Gareffi, G. and Korzeniewicz, M. (Eds.). (1994). *Commodity Chains and Global Capitalism.* Westport, CT: Praeger.

Gordon, A. (1985). *The Evolution of Labor Relations in Japan: Heavy Industry, 1853–1955.* Cambridge: Harvard University Press.

Hamper, B. (1991). *Rivethead.* New York: Warner Books.

Harvey, D. (1985). *The Urbanization of Capital.* Baltimore: Johns Hopkins University Press.

_____. (1989). *The Condition of Postmodernity.* London: Basil Blackwell.

Hopkins, T. and Wallerstein, I. (1986). "Commodity Chains in the World Economy Prior to 1800," *Review* 10, (1): pp. 157–70.

Kato, T. and Steven, R. (1993). *Is Japanese Management Post-Fordism?* London: Mado-sha.

Kenney, M. and Florida, R. (1988). "Beyond Mass Production: Production and the Labor Process in Japan," *Politics and Society* 16, (1): pp. 121–58.

_____. (1993). *Beyond Mass Production: The Japanese System and Its Transfer to the United States.* New York: Oxford University Press.

_____. (1994). "Japanese Maquiladoras: Production Organization and Global Commodity Chains," *World Development* 22, (1): pp. 27–44.

Kenney, M.; Romero, J.; and Cho, D. (1994). "Japanese and Korean Investment in the Maquiladoras: What Role in Global Commodity Chains?" Paper prepared for the International Conference on The Maquiladoras in Mexico: Present and Future Prospects of Industrial Development held at the Colegio de la Frontera Norte, Mexico from May 16–18.

Koike, K. (1988). (Translated by M. Saso). *Understanding Industrial Relations in Japan.* New York: St. Martin's Press.

Krafcik, J. (1986). "Learning from NUMMI," MIT International Motor Vehicle Program Internal Working Paper 15, September.

Kyotani, E. (1994). Seminar at Institute of Governmental Affairs, University of California, Davis, February 28.

Lipietz, A. (1987). *Mirages and Miracles.* London: Verso Books.

NUMMI. (1986). "New United Motor Manufacturing Inc. and the United Automobile Workers: Partners in Training," Case Study for the ILO/Turin Center Seminar on Labor-Management Cooperation in Training and Retraining, May 21.

Parker, M. and Slaughter, J. (1988). "Management by Stress," *Technology Review* (October): pp. 36–46.

Perez, C. and Freeman, C. (1985). "Microelectronics, Long Waves and World Structural Change," *World Development* 13, (3): pp. 441–63.

Pine, B. (1993). *Mass Customization: The New Frontier in Business Competition.* Boston: Harvard Business School Press.

Piore, M. and Sabel, C. (1984). *The Second Industrial Divide.* New York: Basic Books.

Reich, R. (1991). *The Work of Nations.* New York: Alfred Knopf, Inc.

Roobeek, A. (1987). "The Crisis of Fordism and the Rise of New Technological Paradigm," *Futures* 19, (2) 129–54.

Sayer, A. (1986). "New Developments in Manufacturing: The Just-In-Time System," *Capital and Class* 30: pp. 42–72.

——— . (1989). "Postfordism in Question," *International Journal of Urban and Regional Research* 13, (4): pp. 666–95

Sayer, A. and Walker, R. (1992). *The New Social Economy.* London: Basil Blackwell.

Sklair, L. (1989). *Assembling for Development.* Boston: Unwin Hyman, Inc.

Templin, N. (1994). "Auto Plants, Hiring Again, Are Demanding Higher-skilled Labor," *Wall Street Journal* (March 11): pp. A1, A8.

Upton, D. (1994). "The Management of Manufacturing Flexibility," *California Management Review* 36, (2): pp. 72–89.

United Auto Workers. (1991). "Agreement Between New United Motor Manufacturing Inc. and the UAW," July 1.

Womack, J.; Jones, D.; and Roos, D. (1990). *The Machine that Changed the World.* New York: Rawson and Associates.

Zuboff, S. (1988). *In the Age of the Smart Machine: the Future of Work and Power.* New York: Basic Books.

4

PREPARING WORKERS AND STUDENTS FOR THE NEW WORKPLACE

Larry Mikulecky
School of Education, Indiana University, Bloomington, IN 47401 USA

BASIC SKILLS DEMANDS OF THE NEW WORKPLACE

Research has provided us with a good deal of information about literacy demands in the workplace. Researchers (Sticht, 1982; Mikulecky, 1982) have consistently found the vast majority of prose material (i.e., memos, manuals, trouble-shooting directions, new product information) in the workplace to be of high school to college level difficulty. These high school to college difficulty levels are comparable to the difficulty levels of most newspapers and magazines (Wheat, Lindberg, and Nauman, 1977). The number of workers being called upon to use print materials is also being influenced by technological and organizational changes in the workplace. As workplaces restructure to become more productive, workers in many manufacturing and service occupations are being called upon to monitor quality by gathering information from charts, graphs, and computer screens, taking measurements, calculating averages, graphing information, entering information onto various forms, and writing brief shift reports indicating problems and attempted solutions. Others must gather information from print to effectively participate in quality assurance groups and to play active roles in improving productivity (Chisman, 1992; Faison et al., 1992).

The U.S. Department of Labor has reviewed changing demands in many industries and described increased workplace basic skill requirements in publications produced by the Labor Secretary's Commission on Achieving Necessary Skills (U.S. Department of Labor, 1991). The U.S. Department of Labor instituted an advisory committee on skills necessary for functioning in today's workplace (Secretary's Commission on Achieving Necessary Skills; hereafter SCANS) in 1990. The charge of this commission was to determine what skills America's schools should be teaching to prepare student's for the current and future workplace. In addition to higher levels of basic reading, writing, computation, and oral communication skills, the Commission reported that increased competencies are needed in the areas of:

- Identifying, organizing, planning, and allocating **resources**.
- **Interpersonal** skills that include teamwork, teaching, client services, leadership, negotiating, and working with diversity
- Acquiring, organizing, interpreting, evaluating, and communicating **information**.
- Understanding complex inter-relationships within **systems** and distinguishing trends, predicting impacts, as well as monitoring and correcting performance.
- Working with a variety of **technologies** and choosing the appropriate tool for the task.

According to SCANS analyses, work tasks require the integration of these skills in increasingly complex mixtures. For example, one might move from simply scheduling one's own time, to scheduling time for a group while using technology to gather and analyze information, to jointly planning and modifying plans for large-scale projects involving the activities of several groups. The computation, communication, interpersonal, and technological skills required for productive participation in such activities is high and beyond the previous experiences of many workers and high school students.

The SCANS findings are supported by earlier research. Workplace uses of literacy and basic skills are usually embedded in multi-step problemsolving applications (Mikulecky and Ehlinger, 1986; Guthrie, 1988). Problems are presented (either orally or in print), notes are made, information is gathered from one or more print or oral sources, information is evaluated (individually and in groups), decisions are made, and results are monitored. This happens across jobs in the manufacturing and service sectors.

Similar demands occur in terms of quantitative skills which are sometimes described as numeracy (i.e., calculation, measurement, handling data, and problem-solving). Usually these aspects of numeracy overlap each other and also overlap with literacy use. It is of some use, however, to separately examine some of the quantitative skills called for in the workplace. Lloyd and Mikulecky (in press) provide an analysis and discussion of some of these uses.

Examples of basic workplace *calculations* include adding up the total price on a delivery form, finding a percentage discount on an insurance premium, subtracting two fractional lengths marked on a blueprint, and working out the proportions for mixing two chemicals. The method used will depend on the calculation and the context in which it is done. A worker will choose (or may be forced into a particular choice by the circumstances) mental arithmetic, paper-and-pencil, or calculator. Ideally, the worker will check the answer to see if it is reasonable using another, approximate, method. This level of precision and cross-checking in choosing appropriate techniques and estimating sensibly is a new basic skill which goes beyond many school classrooms.

Measurement in the workplace can involve not only direct measuring with ruler or micrometer, but also setting dials on machines which then do the measuring. In either case, care and accuracy are required, as well as an appreciation of the permitted tolerances in a particular situation. Examples include measuring lengths of wood with a ruler to an accuracy of sixteenths of an inch, measuring diameters of wire with a micrometer to within thousandths of an inch, and setting a machine to weigh bags of sugar to an accuracy of tenths of an ounce. In all these case, workers need to know how much error is allowable and how much will produce problems. Few classrooms currently address such a variety of skills.

Workplace *data* come in many forms, and so do the methods for handling those data. Much data are now computerized, and so workers need to be able to interpret what they see on a screen and use a keyboard to react to it. Other workplace data appear as printed graphs, charts, or blueprints. For example, a customer service representative dealing with a client's query may have to key in an identity code to bring up on a screen the details of that client's case, and then read off the print and quantitative information relevant to the particular question. This may also involve making calculations and changes to the computer entry as a result of that query. All this must be done while talking to the client on the telephone. In manufacturing, a worker monitoring the production of a machine may receive a graphic printout every hour and need to interpret the information contained in the graph in order to make changes to the machine's operation.

Problemsolving in the workplace will usually bring together some of the other areas mentioned above, but it has its own characteristics. Problemsolving may entail making judgments about the appropriate action to take, or about the relative weight to give different pieces of information, or about the information that is needed to make a decision. Often problemsolving will be a shared activity, requiring communication, teamwork, and analysis of oral or written information in various formats. This may make the decisionmaking less of a burden on one individual, but it involves other skills of cooperation such as explaining and writing clearly, listening carefully and reaching consensus. Because problemsolving interacts so much with the other skill areas, it is difficult to give examples of it in isolation, so it will be considered further in the detailed examples of workplace tasks which follow. Note how several skill areas can be involved in one workplace task.

Example 1

Statistical Process Control is a way of monitoring the production of a machine while it is in operation, and making necessary adjustments in order to maintain a standard quality of production. At regular intervals, the machine operator takes samples of the product (which may be manufactured parts, lengths of planking, or packages of macaroni and cheese) and takes *measures* (e.g., volume, length, or weight) of each unit in the sample. These *data* are then recorded on a form and the mean and the range of the sample values are *calculated*. The worker plots the mean and range on a *graph*, which already includes values from samples taken earlier in the day. Particular care is made in noting when measurements exceed allowable tolerances.

Then comes the *problemsolving*. The worker needs to decide the degree to which production is still within pre-set tolerances and, if not, by how much to adjust the settings on the machine. Too small an adjustment will not solve the problem, whereas too great a change will turn under-weight into over-weight, producing a new problem. Keep in mind that the machine is still in production while all this is happening, and a worker's speed in dealing with the sample and making the decision will determine costly wastage of materials and production time.

Clearly, for workers to be able to keep their machines producing units of usable quality, it is necessary for them to have a thorough grasp of what they are doing. And they must certainly avoid mistakes in measuring, recording, calculating, and graphing, because any such errors will invalidate the whole process. There is currently little in the experiences of most students and many workers to prepare them for understanding and carrying out Statistical Process Control with accuracy and good judgment. Many even lack basic understandings of numerical and graphing procedures prerequisite for training.

Example 2

A customer service representative (CSR) handling billing inquiries receives a telephone request for late payment. After asking for the customer's name and making notes on the nature of the question, the CSR pulls up the whole of that customer's *record* on a computer screen and checks on the payment history. At this point, there is an initial *decision* to be made: How reliable is this customer? Can any extension of time be given?

If a decision to extend payment time is unclear, the CSR then consults a *set of rules* concerning cut points, the length of allowable extensions, and whether some percentage of the bill must be paid immediately. After determining and *calculating* the effect of the rules in this case, the CSR tells the customer the result—and probably generates a discussion on the possibility of the customer paying as required. Again, the CSR must make a *decision*, within the limits of certain discretionary powers, which will produce the solution most likely to lead to payment and a satisfied customer. Such decisions are often made by high school graduates who have not chosen to go on for further education.

Once again, in this example, the worker is processing data, making calculations, and then making decisions, all under the pressure of time. Here, it is the customer on the telephone generating the pressure, rather than a machine which continues to produce.

Example 3

In high-performance workplaces, teams of workers representing activities at various stages of production often meet in what are called "quality assurance teams." The purposes of such teams are to identify problems, jointly set new productivity goals, and discuss the results of monitoring productivity at various stages of production (i.e., Where are the mistakes or slow-downs happening and what can be done about it.?) An expanded set of basic skills is required.

A typical team problem is too much inventory (i.e., skids loaded with parts or finished product) on the floor. A major productivity goal is "just-in-time" production so material is ready for the next stage of production or for the customer exactly when it is needed. This cuts down on spoilage, breakage, pilfering, and needed warehouse space.

To solve problems like this, four to six workers may spread inventory graphs and relevant memos on a table. Line graphs record the amount of inventory in various locations at various points in time (i.e., by the hour, day, week, or month). Memos may indicate problems identified by other workers. Synthesizing information from these graphs and memos can allow the team to determine when build-ups of excess inventory (i.e., parts or finished product) are occurring. Team members able to process this information will offer problemsolving suggestions on why the build-ups are occurring (i.e., the employee at stage 4 is having a hard time keeping up, the second machine at stage 3 has been requisitioned for a special order project, breakdowns are occurring on the first machine at stage 3). Additional information is then gathered on suggested possibilities. For example, what would happen if a worker at stage 3 went to help at stage 4 every other hour? Alternative computations of output might involve computing half-day splits of time or two-hour splits of time. Speculations about machine breakdowns might involve checking when the machine was last overhauled and recalibrated, looking up projected times between maintenance, and computing time to go. Pulling up machine records of the questionable machine during a comparable time during the last maintenance cycle would provide information to justify a call for early maintenance. New and old work orders would be scanned to see how many parts are called for the special order which took a machine off line. Based on performance so far, computations and estimates would be made for how long it would remain off line.

The culmination of all this brainstorming and information analysis would be a provisional plan to increase production speed and reduce the amount of inventory on the floor. This would involve deciding which workers and machines would do what tasks during which times. It would involve setting goals, counting and making measurements at regular intervals, and recording data to monitor the various stages of production. This plan would need to be documented and broken down for workers not present at the meeting. After three to five days time, the quality assurance team would meet again briefly to determine how well goals had been met and how well problems had been solved.

Since workers in restructured workplaces are expected to be more flexible and do each others' jobs, workers are much more likely to encounter relatively unfamiliar information which they need to sometimes manage independently and are much more likely to face new print demands as part of on-going workplace retraining (Reder, 1994). Though media accounts occasionally describe the de-skilling of parts of some jobs, the overall impact for workers in restructured workplaces is the need to perform more tasks and increasingly complex tasks.

BASIC SKILLS LEVELS OF CURRENT EMPLOYEES

To give some idea of the gap between the need for skills in the workplace and the actual skills of workers, consider some of the results of the recent National Adult Literacy Survey or NALS (Kirsch et al., 1994). The more than 26,000 adults who participated in the two-hour survey were given tasks of varying difficulty to accomplish using realistic scenarios and materials. The percentage of adults correctly completing various prose, document, and quantitative tasks was used to establish task difficulty ratings along a 0–500 scale. By examining particular items, it is possible to develop a more concrete sense of what numerical scores mean. For example, an individual with a score below 149 on the prose scale would have a very low probability of being able to locate consistently details like the name of a country in a short newspaper article. Scoring below 170 on the document scale would indicate a high probability of not being able to locate consistently details like the expiration date on a driver's license. Below 191 on the quantitative scale indicates a high probability of not being able to accurately and consistently total a fairly simple set of figures like those on bank deposit entry. Any score between 0–225 is sometimes reported as being in Level 1, the lowest of five proficiency levels on the NALS. Level 2 scores range from 226–275. Items that represent tasks within Level 2 are listed below along with the difficulty scores of the items.

Prose	250	Locate two features of information from sports article.
	275	Interpret instructions from an appliance warranty.
Document	230	Locate an intersection on street map.
	259	Locate and enter background information on an application for a social security card.
Quantitative	238	Calculate postage and fees for certified mail.
	270	Calculate total costs of purchase from an order form.

Most of the discussion in this chapter will revolve around the percentages of workers in various occupational categories who score at Levels 1 and 2. This is because the literacy and computational demands of changing workplaces and of retraining tend to be higher than Level 2. In other words, workers in Levels 1 and 2 are candidates for basic skills instruction if they must train for new higher skill jobs, retrain for evolving current jobs, or simply be able to handle the increasing difficulty of basic skill demands of restructured jobs created by technological and organizational changes.

Among respondents to the NALS, 20,643 individuals answered questions about their previous and current employment. Responses were coded using the US Census Bureau's 1990 Occupational Classification system which includes several hundred occupational categories. Average proficiencies of workers in most occupational areas tend to be at Level 3 (276–325) or the very low end of Level 4 (326–375). The examples below provide some concrete sense of tasks which could be performed accurately and consistently by workers in the top 50–60% of those taking the test.

Prose	288	Write a brief letter explaining a credit card bill error.
	316	Read a lengthy article to identify two behaviors that meet a stated condition.
	328	State in writing an argument made in a lengthy newspaper article
Document	298	Use a sign-out sheet to respond to a call about a resident.
	323	Enter information given into an auto maintenance form.
Quantitative	278	Use a calculator to calculate the difference between regular and sale price from an advertisement.
	321	Calculate miles per gallon using information on a mileage record chart.
	331	Determine correct change using information in a menu.

In many occupations, however, more than half of workers performed below the national average for all occupations. For example, average proficiencies below Level 3 were found for the following job classifications: laborer, service, farming, forestry, fishing, craft, machine operative, and transportation operative. In addition, in many occupations, 20–40% of workers scored at Levels 1 and 2. Most of the workplace basic skill examples provided in previous pages appear comparable to survey items of Level 3 or higher.

The survey information should not be construed to mean that the country is filled with illiterates in the sense of adults unable to read any print. Most of the 1 in 5 adults who demonstrated proficiencies at Level 1, the lowest level of the NALS, could read some simple material, some of the time. The probability estimates are, however, that most would be unable to consistently and independently do tasks similar to the very simplest literacy and computational tasks called for in today's workforce (i.e. locating and entering key verbal and numerical information on documents and forms or accurately setting up and performing simple addition and subtraction calculations). It is entirely possible that when such tasks arise in the workplace, Level 1 adults must find help, make guesses, or simply stop working.

In high-performance workplaces, where heightened training and information processing requirements abound, job demands are roughly comparable to Level 3 tasks on the NALS. Both the 20% of adults in Level 1 and the next quarter of adults who score within the Level 2 category are likely to experience difficulty with Level 3 tasks like interpreting unfamiliar written instructions, performing multi-step calculations, and using graphical displays like maps and complex charts.

For every occupational area, some individuals performed within Level 1, the lowest level. The percentages range from 1–3% of teachers and other professionals, to some very high percentages in the health support occupations (13–8%). Approximately 10% of sales personnel also perform at the lowest level while the percentages of clerks and secretaries at the lowest level range from 5– 8%.

More to the point is the fact that for occupations calling for a good deal of paperwork and information processing (i.e., health support, sales, clerical, and secretarial), many workers seem unable to comprehend moderately simple materials (i.e., Levels 1 and 2). Approximately 40% of health support personnel and sales personnel perform at Levels 1 or 2 while approximately 30% of clerks and secretaries perform at these levels. Again, performing in Level 2 does not mean that one is illiterate. In fact, over 90% of adults at this level did report themselves as having literacy difficulty. It does mean, however, that being able to independently and consistently comprehend most training and current information is likely to be very difficult.

CURRENT SCHOOL PRACTICES AND STUDENT ABILITIES

Though there have been many experiments at school reform throughout the 1980s and 1990s, National Assessment of Educational Progress or NAEP data reveal only small changes in the practices and abilities of students. Students do not expend much school time or homework time with reading, writing, or calculation and average abilities have remained about the same while external demands in the workplace have increased.

School Practices

Comparisons of school and workplace literacy (Mikulecky, 1982) indicate that students average less time reading in school and doing homework (98 minutes daily) than do workers on the job (2 to 3 hours daily). The school average is somewhat deceptive, since about 25% of students read several hours daily while most read well below an hour daily and a third less than 20 minutes daily. In addition, the textbook reading to answer simple factual questions performed most often by students is less complex than the types of literacy tasks faced by many workers.

The data indicating that most high school students perform little reading and writing related to class work have been consistent for more than a decade since the 1980s. Applebee (1981) reported that only 3% of student writing was as long as a paragraph, and the majority of writing was in response to text or worksheet questions. The majority of information in classrooms was distributed through lecture or teacher-dominated discussion, and what reading did occur was primarily for the purpose of answering short worksheet or end of chapter questions. Goodlad's (1983) observations of over a thousand classrooms indicated that over 70% of student time in the 1980s was spent listening in classrooms and the vast majority of this listening was to teacher talk. There was very little time for practicing reading, writing, or team skills.

The NAEP regularly tests and surveys the abilities and study habits of a cross-section of the nation's students. Some items on the NAEP ask students about daily reading times and amounts. There are few differences in times and amounts reported by 17-year-olds in the '80s and '90s (Langer et al., 1990). School reform during the 1980s did not appear to have a major impact on the average amount of reading performed by students.

Some teachers hopefully suggest that literacy practice occurs at home during homework. Recent results for the NAEP survey of adolescent homework patterns suggests this is misplaced hope for most students (Langer et al., 1990). Seventy percent of high school and middle school students do less than 2 hours of homework daily.

Nearly 30% of eighth graders do less than one-half hour of homework daily and by grade 12, nearly 40% of students do less than one-half hour of homework daily. For many, 12th grade means less work than 8th grade. The NAEP also asked about the amount of homework completed each day. It is not very much. For all of their classes combined, 61% of eighth graders report doing less than 10 pages of homework each day. More than 30% do less than 5 pages. These figures remain essentially the same for 12th graders. To improve a grade level in reading, writing, and computational abilities takes several hundred hours of practice. The majority of students currently aren't getting that practice either in school or while doing homework. To be fair to schools and teachers, the amount of homework assigned has increased. It is not, however, being done.

Student Abilities

A reasonable way to judge the effectiveness of recent education is to assess the abilities of students in school as well as the abilities of young adults who have recently matriculated through school. The NAEP has performed several national studies of adolescent reading abilities since the early 1970s and surveyed young adult (21–25 years of age) literacy abilities in the mid-1980s (Mullis and Jenkins, 1990; Kirsch and Jungeblut, 1986). NAEP study authors report that the overall reading performance of 17-year-olds improved somewhat between 1971 and 1988, mainly in lower level skills. Nearly 100% of this age group is now able to read at a *basic* level. According to NAEP test items, achieving at the *basic* level implies being able to:

- Follow brief written directions.
- Select words, phrases, or sentences to describe a simple picture.
- Interpret simple written clues to identify objects.
- Locate and identify facts from simple informational paragraphs, stories and news articles.
- Combine ideas and make inferences based on short, uncomplicated passages.

In addition, in 1988, 86% of 17-year-olds were able to perform at the *intermediate* level. This was an improvement from a generation earlier when only 78% of 17-year-olds attained this level. According to NAEP, performance at the *intermediate* level implies being able to do all the tasks described above in addition to being able to:

- Search for, locate, and organize information found in relatively lengthy passages.
- Recognize paraphrases of what has been read.
- Make inferences and reach generalizations about main ideas and author's purpose from passages dealing with literature, science, and social studies.

To place this *intermediate* level in perspective, it should be noted that this is the level achieved by the average 13-year-old or eighth-grade student. The majority of school assignments described by earlier researchers appear to involve simple reading, paraphrasing, and locating key ideas in textbook passages. Thus, it appears that schools and teachers have succeeded, for the most part, in helping the majority of students to learn the simple skills being taught. It also means that an average classroom of 17-year-olds still has four or five students unable to recognize paraphrases and draw inferences. Little in the school assessments relates directly to the sorts of tasks described earlier in workplace examples.

Greater difficulties in student performance begin to emerge on the more difficult items. Only about 42% of 17-year-olds reach the *adept* level. Student attainment of this level has remained essentially the same since 1971. Readers who achieve at the *adept* level are able to perform at the *intermediate* levels and:

- Understand complicated literary and informational passages.
- Analyze and integrate less familiar material and provide reactions to and explanations of the text as a whole.

These tasks are somewhat similar to demands in many workplace training classes involving texts and manuals.

Only 4.8% of 17-year-olds in 1988 performed beyond the *adept* level and achieved at the *advanced* level. This percentage of students achieving at the *advanced* level has dropped significantly since 1971 when 6.6% achieved at this level. At the *advanced* level, readers are able to perform at the *adept* level and:

- Extend and restructure the ideas presented in specialized and complex texts. Examples include scientific materials, literary essays, historical documents, and professional material .
- Understand links between ideas even when those links are not explicitly stated.
- Make appropriate generalizations even when texts lack clear introductions and explanations.

Slightly more students today can manage with simple skills, which are seldom called for in the workplace, than was the case previously. It appears that the cost of this gain in basic skills has been that a slightly lower percentage of students are now able to perform complex tasks similar to those called for in colleges and more challenging jobs.

Young Adult Abilities

In 1986, NAEP released a major study of the functional literacy abilities of 21- to 25-year-old young adults (Kirsch and Jungeblut, 1986). A few observations about the results are in order. When viewing percentages of the total 21- to 25-year-old population, it appears clear that there is not a large degree of basic "illiteracy." Over 95% of young adults could:

- Sign their names.
- Locate expiration dates on a driver's license.
- Locate a time on a meeting form.
- Enter a caller's number on a phone message form.

- Write about a job they would like.
- Locate a movie in a TV listing.
- Enter personal information on a job application.

As with the NAEP data on adolescents, wide racial and ethnic differences appear in the young adult data. Though a vast majority of all ethnic populations can accomplish basic literacy tasks, gaps in populations become even wider as the complexity of tasks increase. For example, according to NAEP data it is probable that 22% of whites would have difficulty with consistently and accurately perform tasks like writing a letter to state that an error was made in billing. On the same item, 60% of blacks and 42% of Hispanics would be likely to have difficulty. Test data indicate that it is probable that 35% of whites would have difficulty with consistent and accurate performance following directions to travel from one location to another using a map. On the same item, 80% of blacks and 63% of Hispanics would be likely to have difficulty.

The young adult literacy data make several points very clearly. The vast majority of young adult Americans have mastered basic literacy demands. This vast majority shrinks rapidly as literacy tasks increase to moderate complexity. Finally, differences in performance between whites, blacks, and Hispanics are extremely wide and distressing, especially at middle and upper levels of the test. It appears that substantial proportions of our young adult population, and especially our minority population are ill-equipped for the high and increasing literacy challenges associated with being productive and self-sufficient in our society.

PROMISING DEVELOPMENTS IN SCHOOL AND WORKPLACE EDUCATION

In relation to new workplace demands, there have been several promising developments in education over the past few years. These include a widespread trend toward having students function more in cooperative groups and teams at all grade levels as well as broad range planning of school-to-work transition programs such as Tech Prep and Career Academies at the high school level. These school developments are far from having a major impact, but do indicate increased perception of and response to workplace demands by educators and school boards. In addition, basic skills training of adults in the workplace has expanded greatly, though only to about 20% of workplaces and less than 10% of adults.

Cooperative Learning

During the past decade, the use and research of cooperative learning in classrooms has expanded to the point where data bases include more than 1,000 research references. Forms of cooperative learning are used in classrooms from elementary school to university level. Robert Slavin, as well as David and Roger Johnson, are recognized leaders in developing and researching concepts related to cooperative learning. By the early 1990s, syntheses of research demonstrated that cooperative learning methods using group rewards and individual accountability consistently increased student achievement more than control methods (Slavin, 1991). Johnson and Johnson (1990), key researchers in cooperative learning for more than a decade, had established that people do not know instinctively how to interact effectively with others. For cooperation to succeed, students must get to know and trust one another, learn to communicate accurately and unambiguously, accept and support one another, and resolve conflicts constructively. These skills can be taught and learned by students. Educational journals and teacher training sessions regularly offer suggestions and plans for developing cooperative learning activities in classrooms. Though no study reports the degree to which teachers are employing these group and team project approaches nationally, most educators find it rare to enter a school in the 1990s in which some teachers do not use such techniques at least occasionally.

Tech Prep

Tech Prep is a trend more directly related to helping students prepare for the new workplace. In the early 1990s in the U.S., the Carl D. Perkins Vocational and Applied Technology Act funded programs to help 16- to 20-year-olds complete high school and then go on to post-secondary education in a technical or practical field requiring an Associate's degree or Bachelor's four-year degree. These programs were popularly entitled, "Tech Prep." By 1994, versions of Tech Prep programs were in operation or being considered in the United States, Canada, Mexico, the United Kingdom, Brazil, and Turkey. Programs embrace the following elements:

- Partnerships between educators and local employers.
- Integration of academic skills, vocational skills, and basic skills.
- Local control in developing curricula and assessment methods to help high school youth to make the transition to college demands and the new workplace.

In the United States, funding is through federal money allocated to local consortia of secondary schools, post-secondary schools, and employers in each of the 50 states. In 1992 there were 850 consortia and by 1994 over 1,200 consortia. An average consortium includes 12 secondary schools, 2 post-secondary schools, and 10 employers. Nearly 50% of secondary schools and a higher percentage of two-year post-secondary institutions participate in Tech Prep curriculum reforms.

According to Bragg et al. (1994), program coordinators vary somewhat in describing whom they see as a target audience for Tech Prep. Most report targeting students in the middle-half, though about 11% see Tech Prep as being for all students, and another 14% see the target group as being lower ability students. This variation is because federal guidelines allocate most decisions to local control. In 1994, the predominant Tech Prep curricular reform (70–90% of secondary schools) involved supplementing vocational and/or academic classes with some type of commercial or locally developed materials which apply skills to workplace examples. Lower percentages of schools have been able to implement interdisciplinary programs and work-based learning, though these are most apparent in programs which have been in existence for two or more years. It is difficult to determine the number of students participating in Tech Prep programs since most programs begin by bringing together employers, secondary teachers, and post-secondary teachers to do local planning. A report on student involvement in the state of North Carolina provides reason for being cautious about the degree of actual student involvement (North Carolina, 1993). In North Carolina, several types of school-to-work transition programs have been surveyed. Nearly 100,000 students are served annually by a variety of programs within the state. Less than 3,000 of these students were involved in Tech Prep programs in 1993. Other students participated in less complex school-to-work experiences like cooperative education (14,000 students), internships (2,000 students), and vocational student organizations (80,000 students).

Even though the number of students involved in Tech Prep programs is relatively small, Tech Prep is still arguably the school reform initiative to have had the most widespread impact in secondary schools. It has provided a framework for teachers from different subject areas and levels to meet with employers to address local needs and plan cooperative ventures. To some degree, it is now present in most school districts across the country. Though it has been slow in getting beyond the planning stage, some form of Tech Prep is likely to survive in future school-to-work transition programs at the federal and state levels. Newly funded school-to-work initiatives in the United States appear likely to give even greater support for local collaborative ventures involving employers and educators in secondary as well as post-secondary schools.

Career Academies

A more ambitious (though even less accessible) form of school-to-work transition is the experimental career academy which now operates in a limited number of select sites. Career academies are special programs within high schools which focus upon an industry or broad family of occupations and coordinate three to four years of education to link academic training to the skill demands of targeted career areas. Burnett (1992) reports that most career academies share a number of attributes:

> school-within-a-school format, voluntary student participation, broadly-defined career themes, integration of academic and vocational curricula, scheduling that keeps students together throughout the day and throughout the three- or four-year program, rigorous academic courses, work experience, high levels of involvement by local businesses, and outside funding from business and government sources.

The academy model coordinates curriculum across several years of education, integrates academic and vocational courses and fosters collaboration between teachers. Most academies link class themes, large-scale projects, and other applications of skills to the cooperating industry (i.e., electronics, health care, space technology). Career academies are predominantly in urban districts and have served disadvantaged populations at risk of dropping out. Faculty and staff participation is voluntary and often teachers are specially recruited. Career academies require a significant financial investment by school districts. Evaluation of program impact is on-going, but appears to be positive in the short-term and promising in the long term. Fewer than 20 career academies were in operation by the mid-1990s.

Workplace Basic Skills Education

For more than a decade, basic skills courses have been offered by partnerships between the United Auto Workers and automotive manufacturers such as Ford, General Motors, and Chrysler. Contract negotiations create separate training agencies with independent budgets which are jointly managed by a board of union and management representation. Training is for union members and, sometimes, spouses. It can address any training focus requested by members. Total training budgets for these joint ventures are several hundred million dollars per year. In the late 1980s, the Ford/UAW partnership, alone, was budgeted at over $80 million per year. It is not clear what proportion of these funds are allocated to basic skills and workplace literacy training. Many union/management agreements articulate guidelines which clearly state that management will pay for job and technical training while the joint ventures pay for self-development and employee benefit training, including basic skills education.

The automotive industry is not the only industry providing workplace basic skills instruction. American employers spend vastly more upon basic skills instruction than do taxpayers. Though no clear-cut study of funding sources for US workplace basic skills programs exists, the American Society for Training and Development surveyed 20,000 U.S. employers with more than 100 employees to determine training expenditures. The survey estimates that American business spent more than $44.4 billion for training in 1989. A small fraction (less than 10%) of this went to remedial education. Still, this is likely to be more than $4 billion spent on remedial training. About a fifth of companies (22%) reported offering such training. Not all workplace literacy programs are conceived of as remedial training, however. It is not possible to determine what proportion of workplace literacy training is subsumed under categories like communication skills, technical skills, clerical skills, team-building, problemsolving, and several other categories listed in the survey (Oberle, Gerber, and Gordon, 1989). The amount of workplace literacy training subsumed under these labels is likely to be high, however, since the general trend has been for U.S. businesses to abandon remedial terms like "literacy training" in favor of terms like "technical preparation," and "communication skills". In any case, the degree to which employers and unions are funding workplace literacy training vastly exceeds the approximately $20 million per year which the federal government has allocated to the National Workplace Literacy Program. The $4 billion or more expended by employers is vastly greater than the $300–$400 million per year allocated by the federal government to the full range of basic skills programs for all adults.

Increasingly, consortia of employers and unions have begun to cooperate to provide information and workplace basic skill services. The United Steel Workers of America, working together with a dozen major steel corporations, has created the Institute for Career Development. Among the projects taken on by the Institute for Career Development is a major basic skills task analysis of job families in the steel industry as a prelude to developing custom-designed basic skills instruction related to daily demands faced by union members. Similarly, in the electronics industry, the Electronics Industries Association has surveyed members and set up a task force to explore the possibility of competitors within the electronics industry cooperating to develop workplace linked basic skills instruction for employees from within all cooperating companies.

Though the past decade has seen major advances in basic skills instruction in the workplace and impact has been great in some workplaces, national impact is still quite limited. Adults participating in the National Adult Literacy Survey were asked if they had ever participated in basic skills education. Of employed adults, less than 10% indicated having received such training. Most adults who will hold jobs over the next 20 years are already in the workplace. Given the gaps between skill demands and worker skill levels, the fact that only 10% have participated in basic skills training suggests there is a long way to go (Mikulecky, in press).

SUMMARY

Basic skill demands in the workplace have both expanded and increased, especially in restructured workplaces where technology and reorganized work require more flexibility, cooperation, and independent competence of employees. The most recent National Adult Literacy Survey indicates that a very large percentage of adults do not currently posses these skills. Though workers, on average, appear to have the capabilities to perform the complex tasks called for in the new workplace, nearly every occupation area has a significant fraction of current employees who cannot consistently and accurately perform at expected levels. In several manufacturing and service sector occupational classifications, more than half of workers were unable to consistently perform literacy and quantitative tasks similar to job tasks in restructured workplaces.

Over the past 15 years, school tests have shown little change in student literacy abilities or the amount of reading and homework students perform daily. There has been a slight improvement in the percentage of students able to demonstrate low level skills and a slight decrease in the percentage able to demonstrate higher level skills. Few students of any sort are able to perform at high levels, but percentages are particularly alarming among students from minority and low-income homes.

There are indications that schools and employers are beginning to recognize and respond to the gaps between the basic skill demands of the new workplace and the current ability levels of adults and adolescents. The new programs appear promising and on the increase, but the actual impact on adolescents and adults has been very limited, to date.

REFERENCES

Applebee, A. (1981). *Writing in the Secondary School.* Urbana, IL: National Council of the Teachers of English.

Bragg, D. et al. (1994). *Tech Prep Implementation in the United States: Promising Trends and Lingering Challenges*. Technical Report. Macomb, IL: National Center for Research on Vocational Education, Western Illinois University.

Burnett, G. (1992). *Career Academies: Educating Urban Students for Career Success.* ERIC/CUE Digest, Number 84. ERIC Clearinghouse on Urban Education, New York, NY (ED355311).

Chisman, F. (1992). *The Missing Link: Workplace Education for Small Businesses.* Washington, D.C.: The Southport Institute for Policy Analysis.

Faison, T. et al. (1992). *Ahead of the Curve: Basic Skills in Four Exceptional Firms.* Washington, D.C.: The Southport Institute for Policy Analysis.

Goodlad, J. (1983). *A Place Called School.* New York: McGraw-Hill.

Guthrie, J. (1988). "Locating Information in Documents: Examination of a Cognitive Model," *Reading Research Quarterly*, 23, pp. 178–99.

Johnson, D.W. and Johnson, R.T. (1990). "Social Skills for Successful Group Work," *Educational Leadership*, vol. 47, no. 4, pp. 29–33.

Kirsch, I. and Jungeblut, A. (1986). *Literacy: Profiles of America's Young Adults.* Princeton, NJ: National Assessment of Educational Progress at Educational Testing Service.

Kirsch, I. et al. (1994). *Adult Literacy in America.* Princeton, NJ: National Assessment of Educational Progress at Educational Testing Service.

Langer, J. et al. (1990). *Learning toRead in Our Nation's Schools.* Princeton, NJ: National Assessment of Educational Progress, Educational Testing Service.

Lloyd, P. and Mikulecky, L. (in press). "Numeracy Skills for Workplace Needs." In I. Gal (Ed.), *National Center on Adult Literacy Guide for Adult Numeracy Educators*. Philadelphia, PA: National Center on Adult Literacy. University of Pennsylvania.

Mikulecky, L. (1982). "Job Literacy: The Relationship between School Preparation and Workplace Actuality," *Reading Research Quarterly*, 17, pp. 400–19.

——. (in press). "Education for the Workplace." In C. Kaestle et al. (Eds.), *Adult Literacy and Education in America*. Princeton, NJ: Educational Testing Service.

Mikulecky, L. and Ehlinger, J. (1986). "The Influence of Metacognitive Aspects of Literacy on Job Performance of Electronics Technicians." *Journal of Reading Behavior,* vol. 18, no. 1, pp. 41–62.

Mullis, I. and Jenkins, L. (1990). *The Reading Report Card, 1971–1988.* Princeton, NJ: National Assessment of Educational Progress, Educational Testing Service.

North Carolina State Department of Public Instruction (1993). *School-to-Work Transition Programs: A Status Report.* Raleigh. NC: Division of Vocational and Technical Education Services (ED364703).

Oberle, J.; Gerber, P.; and Gordon, J. (1989). "Industry Report," *Training*, Minneapolis, Minnesota, Lakewood Publications, Inc. 26 (10), pp. 31–76.

Reder, S. (1994). *Learning to Earn: The Reward of Workplace Literacy.* Philadelphia, PA: National Center on Adult Literacy. University of Pennsylvania.

Slavin, R.E. (1991). "Synthesis of Research of Cooperative Learning," *Educational-Leadership*; 48:5, pp. 71–82.

Sticht, T. (1982). *Basic Skills in Defense.* Alexandria, VA: Human Resources Research Organization.

U.S. Department of Labor. (1991). *What Work Requires of Schools.* Washington, DC.: The Secretary's Commission on Achieving Necessary Skills.

Wheat, T.; Lindberg, M.; and Nauman, M. (1977). "An Exploratory Investigation of Newspaper Readability," *Illinois Reading Council Journal, 5*, pp. 4–7.

5

LARGE FIRM STRATEGIES: SPATIAL PATTERNS OF PRODUCTION

Claire Ellen Pavlik
Department of Geography, The University of Iowa, Iowa City, IA 52242 USA

INTRODUCTION

Views of national economic development and firm evolution that predominated through the early 1970s emphasized stable sequences of developmental stages through which leading nations or firms had passed. Less industrialized nations, if successful, were expected to pass through these same stages. Because the technical and social infrastructure associated with each developmental stage were known in the case of the lead countries, "followers" might be able to decrease the length of each individual stage, thus climbing the development ladder more quickly. For firms, a similar evolutionary model of industrial organization dominated. In this case, however, the description of development focused primarily on leading firms that grew to dominate industrial branches or sectors. In the case of firms, conditions in the market were expected to operate such that the "best" firms developed through a sequence, culminating in a particular organizational form. Firms with inadequate resources, less able management, or poor competitive positions, on the other hand, were expected to remain at lower stages in the sequence unless conditions changed.

More recent views, in contrast, focus less on particular paths of development and more on the way in which interactions affect growth and change. Analyses of the interaction between developed and less developed nations, for example, have led to theories of underdevelopment, in which paths to economic growth and rising standards of living are in part blocked by patterns of pricing, trade, and ownership of resources. For developed nations, newer perspectives focus on regulatory institutions and practices, and the way in which social consensus, or the lack thereof, impacts the workings of and interactions between the economy and society. For firms as well, recent discussions have focused on context and interactions as important factors in determining the relative success of individual firms and regional production complexes.

VIEWS OF NATIONAL ECONOMIC DEVELOPMENT

One of the dominant linear evolutionary models of economic development was put forward by Rostow (1960) in the late 1950s. Based on the evaluation of historical development of industrialized countries, it argued that in the development process, a nation passed through a series of stages. The beginning stage consists of subsistence agriculture and small scale resource exploitation with little commercial activity. The second stage involves a period in which social change and physical investments create favorable conditions for economic development. When conditions are favorable for further development, a third "takeoff" stage can be attained, in which rising investment levels and economic growth, through transformation to a commercially oriented society, occur. In the fourth stage, the economy matures as the technological level of industries in more advanced nations is attained as diversification leads to self-sustaining growth. Once the economy has matured, a period of mass consumption is entered. Since the economy has diversified and investments in production equipment have led to high productivity, consumption of goods and services rise as time spent in formal work, both during the week or year or over an individual's lifetime, fall. Following this fifth stage, some advanced economies have entered a post-industrial stage, in which services replace manufacturing as generators of wealth and employment and information becomes the key resource (Bell, 1973).

Linear evolutionary models have tended to focus on identifying particular economic or social variables that can be used as indicators of the stage of a nation's development and on determining whether the "conditions" at a given time are consonant with those associated with past paths of development. In particular, these models have been employed to try to identify the social or investment conditions that prevent economic "takeoff" from occurring. Because of their focus on monitoring individual nation development, interactions between nations as well as between groups within nations tend to be neglected. In addition, the use of western nations as the basis for the model has produced the image of a homogeneous developmental path. Differences between nations and the institutions and practices that affect social conditions, such as the distribution of income or wealth within nations at similar stages of development, have not been addressed.

Because these models are based on historical cases, they are also of limited use in analyzing the current difficulties faced by developed economies. They provide a good description of the developmental path to date, but do not provide a deep explanation of how the economic problems and crises faced now or in the past developed and could be resolved.

The French "Regulation" School of economic analysis has advanced a different position on the evaluation of economic development of developed countries (Boyer, 1990; Noel, 1987). Rather than take the historical sequence of industrial or social development as necessary or given, economic and technological changes are viewed as a social process and social change is seen as the outcome of conflicts and compromises between sub-national groups. As in more traditional views, the Regulationist perspective tends to focus on nations as the basic "containers" of development. However, while both focus on the development of the same group of western countries, Regulationists see the stages through which these nations passed less as determinate and "natural" steps in a predictable linear progression and more as periods of intermittent social compromise and stability alternating with periods of social upheaval and the institutional restructuring. Stable periods are not the expected and unproblematic development outcomes of "economic progress" but are the result of social alliances, compromises, and "experiments" within societies that lead to a new phase of social peace and economic stability.

During a period of relative economic stability, Regulationists suggest that a particular mode of development dominates. In such a period, two basic conditions hold. First, there is general social adherence to a set of beliefs or regulations under which the members of a society operate: people generally believe that the system works and operate within its rules. Second, production and consumption must balance so that the cycle of economic accumulation is closed: problems of under-consumption and over-production must be minimized. That is, the distribution of production is such that a stable growth path can be maintained. A mode of development can be decomposed into three components: a regime of accumulation, modes of regulation, and a labor process model. The regime of accumulation is the logic or laws at the macroeconomic level that govern the development and social use of production. Modes of regulation are mechanisms that adjust individual behavior to the collective regime of accumulation's principles. The labor process model describes the organization of the labor process within and between firms. When a national economy follows a stable path, these three components are in harmony such that wage formation, industrial relations, and the distribution of income to labor at the microeconomic level (parts of the labor process model) are in accord with the prevailing macroeconomic logic operating at the national level. During a period of economic instability or crisis, either the national economic and social system has changed such that these three components are no longer acting in concert, or stresses such as war, technological change, or instability at the international level have altered the framing conditions within which the system operates so that stable growth is no longer possible.

The instability and uncertainty of the 1930s through the mid-1940s and the period from the early 1970s to the present are periods of economic crisis and restructuring in the Regulationist view. The instability beginning the 1930s was marked by high unemployment, business failures, the restructuring of politics and labor organization, and by the violence of war as social experiments in Italy and Germany (and Japan) were extended by force to other nations. The period that followed, from the late 1940s through the late 1960s, was a period in which the U.S., Canada, and most western European countries pursued similar Keynesian macroeconomic policies and developed welfare state support, exhibited similar conditions for organized labor, and operated within increasingly harmonized rules for commercial activities and industrial relations (e.g., GATT agreements, the European Economic Community).

The early 1970s signaled a less severe breakdown of economic growth and social order than the 1930s. The pattern of high economic growth characteristic of the 1950s and 1960s shifted first to stagnation, then to slow growth. With declining growth rates, there has been erosion of employment as the rate of job formation for jobs with mid-to-high wages has fallen, production and profit rates in many sectors have declined, and the ability of firms to generate new investments has fallen. While new industries such as micro-electronics have developed, these have generated insufficient numbers of jobs to replace those lost in other sectors. Excess industrial capacity has led to plant closure in many traditional industrial sectors, particularly those with lower cost foreign competition, and international instability in exchange rates, commodity and energy costs, and interest rates has added to the difficulties facing firms. For nations, this has led to a reconsideration of the existing status quo in economic and social policies, leading to calls for the end of the Keynesian welfare state. The apparent inability of nations to control exchange rates, levels of public indebtedness, interest rates, and trade balances has resulted in erosion of belief in state management, and in some cases has led to the implementation of liberalist policies including the privatization of state ownership, reductions in welfare transfer payments, and the transfer from state hands to private firms of administrative and state service tasks. Pressure to liberalize labor markets through changes to labor legislation has also developed, and new shop floor practices have begun to reshape the nature of the workplace, at least in some industries.

While these changes can be pointed to as part of the search for elements of a new phase of economic and social stability, there is no clear "solution" to the problems that industrialized nations face. In Regulationist terms, a new mode of development is not evident. While the general contours of the problems faced are similar, individual nations continue to pursue somewhat different paths in the search for higher rates of economic and employment growth. In western Europe, the debate begun in the 1980s (see Boyer, 1988) on how flexibility of the labor force might be increased continues, with the United Kingdom leading the way in breaking down labor market rigidities, increasing employers options for part-time work and numerical flexibility, and decreasing the power of unions, while Germany and France attempt to construct alternatives that allow the current status of the employed labor force to be maintained while generating job growth for the unemployed.

The search for a new model has been difficult because no nation clearly has the role of "leader" as the U.S. had through the mid-1960s. Some view the relatively deregulated labor markets of the U.S. and UK as advisable because of the numbers of new jobs that have been generated. Others point out that low rates of productivity and wages will not meet the twin goals of the maintenance of standards of living and the return to higher growth rates. In the eyes of others, the ability of nations such as Germany to continue to be successful in international markets and retain high wage industrial employment with job security puts them forward as models to emulate. However, while the levels of institutional commitment to the labor force may be instrumental in maintaining high productivity growth, the size of the German labor force has stagnated since the 1970s, leaving new entrants with few employment prospects (Schettkat, 1992). Thus unless nations develop or strengthen export sectors to support domestic employment, as in the case of Japan, there appear to be two current choices for national adjustment. The liberal alternative involves increased institutional flexibility in labor markets, accompanied by higher rates of job formation but lower productivity. The "corporatist" alternative, on the other, involves high commitments to labor in the form of job security, social security expenditures, and unemployment benefits. This can be accompanied by high rates of productivity increase, as long as employers benefit from workers' experience and additional training leads to positive returns.

VIEWS OF FIRM EVOLUTION

Just as descriptions of economic development have focused on leading nations, discussions of the evolution of firms has emphasized firms and industries that seem to lead the way to new forms of internal organization and larger spatial extents of production and marketing networks. In general, the trajectory of development for an industry is seen as one that passes from an early stage with predominately small firms, some of which grow in size and complexity. As markets develop, trade barriers between nations are reduced, the division of labor is refined, and technological progress increases output levels. Ultimately, the largest firms may grow to be large, multinational, multidivisional enterprises. In sectors in which this has not occurred, discussions tend to focus on the industrial, market, or labor peculiarities which tend to block what has become the expected path to "well-developed" industrial organization (see, for example, Scherer and Ross, 1990). Because the U.S. led in the growth of large firms with national or international production and marketing, analyses of the industrial organization of other nations has often focused on identifying the impediments to firm growth that seem to have hindered firms from "catching up with the leaders" to become international competitors (see Jacquemin and de Jong, 1977).

The development of the multidivisional corporate form has been chronicled by business historians, such as Chandler, who have attempted to identify the concurrent developments that led to the formation of new types of internal organization and to commercial success. In his earlier work (1962, 1977) and in the comparison of firms in the U.S., Britain, and Germany (1990), Chandler argues that technological advances in manufacturing processes, transportation, and communication, combined with implementation of a new division of managerial labor and development of marketing expertise led firms organized in the modern corporate form to outpace firms with less sophisticated organizational hierarchies. In Chandler's language, the firm's strategy, combined with a new organizational structure, allowed it to develop production and marketing of a scale and integrated operations of a scope not exhibited before in business history. Others, such as Dunning (1979, 1993), have incorporated these arguments into explanations of the development and success of multinational firms.

The analysis of geographical patterns of production associated with the logic of corporate evolution from groups of small firms to large corporations has focused specifically on the locational traits associated with the expansion of production (during the 1970s and 1980s corporate restructuring was also examined) and on explaining the general patterns of geographical evolution as leading firms expanded. While some discussions focused on the evolution of production patterns for regional and national firms, leading firms with international investments in marketing and production were of intense interest. The two locational models stand out in these discussions of the evolution of international production: the product cycle model and the new international division of labor.

The product cycle model, introduced by Vernon (1966), describes the logic of firm evolution within an international context of the U.S. as the leading economy, with higher levels of labor training, wages, and disposable income, a second group of nations with lower levels of labor training, wage rates, and disposable income, and less developed countries with yet lower levels. The model's focus is on the spatial evolution of production for an individual good. The development of first domestic then international markets is used to trace out the shifting locations of production. Because the initial high cost of research and development (R&D) results in high market prices, it is argued that innovations will tend to arise and be marketed first in nations with the highest levels of disposable income, hence of expenditure on luxuries. Over time, new production locations are developed as production technology is refined so that production costs fall, allowing entry into newer markets with lower levels of disposable income. In addition, as markets which were developed early in the cycle become saturated, the older production facilities that originally served those markets are closed (see Table 5–1).

During the innovation phase, production is expected to be concentrated near the R&D center at which the new product was developed. If marketing trials are successful, the firm extends its marketing effort nationally and develops additional production facilities to serve the national market. These are expected to be located to serve the national market at reasonable cost. As markets within the home country become fully developed, the firm seeks export markets, first in countries with higher levels of income. During the third phase, domestic production and consumption continue to increase as foreign markets expand. Additional facilities are developed to serve these external markets. The fourth phase is marked by price competition from rival firms. In this phase, competition leads to the rationalization of production, as techniques are refined to reduce costs, and of production locations as older and higher cost facilities, often within the home country, are closed. As a result, the home country generally becomes a net importer of the product. The fifth phase, obsolescence, is marked by declining consumption as market saturation occurs or new products are marketed which displace the old. Only the least cost locations remain as production sites, and these may be

Table 5–1. Stages of the Product Cycle

Stage	*Description*	*Production Locations*
1. Innovation	RD based on development of new product; pilot/test production and marketing	Economically developed nation, near company's HQ location
2. Initial marketing	Production and marketing developed in home country; beginning of exports; some foreign production	Company's home country; other economically developed nations
3. Maturity	Expansion of domestic and foreign production; extension of markets	Tendency to locate in or near markets
4. Standardization	Competition leads to production rationalization; developed nations become net importers	Newer market and lower cost locations
5. Obsolescence	Decline in consumption in developed nations; competition forces additional rationalization	Least cost locations; closure of high cost plants

Source: After Vernon, 1966.

locations in less developed nations where wage rates are low and labor training is sufficient to carry out the production process.

The general framework of international conditions on which the product cycle is constructed is historically bounded by the tiered levels of income, wages, and labor training that it portrays. This hierarchical view of nations may have been appropriate for the early post-war years. However, the relative standings of the U.S., western Europe, and Japan have changed substantially. In the early version of the product cycle (Vernon, 1966), it was also suggested that barriers to entry of goods, in particular, would lead to investment in foreign production facilities. Changes in the international regulation of the flows of goods and services, through GATT, bilateral trade agreements, and regional economic areas have altered these framing conditions. In addition, the time frame within which the spatial pattern of production is expected to occur is relatively long. Many, including Vernon (1979), have suggested that the innovative cycles for product niches have been shortened to such an extent that the locational shifts described in the cycle simply can no longer occur in many industries. Along with shorter product cycles, it has been argued that while a particular product may no longer be produced by a firm, the investments firms make in production facilities and equipment are generally not eliminated quickly. Managerial expertise and economies of scope often mean that facilities are multipurpose and the production palettes at those facilities are broader than the product cycle describes. Thus, rather than conceptualize locational changes of a single product within a firm, a more general area of production expertise, often referred to as a "core competence," should be the focus (Foss, 1993). Finally, general technological innovations are said to offer firms the capacity to rationalize facilities and improve products *without* shifting production locations. Automation and the introduction of new production equipment may afford better options to firms than the development of facilities elsewhere (Schoenberger, 1988).

Unlike the product cycle, which focuses on changing locations over time, the new international division of labor (NIDL) position outlines a locational strategy that is expected to apply during a particular phase of international development (see Fröbel, Heinrichs, and Kreye, 1980). As in the product cycle, the world's economy is described as a hierarchy in which developed economies have high wages and standards of living and hold expertise in research and new product development, while the remaining nations vary in their innovative capacity, wage rates, and standards of living. A second framing condition is that the world economy is expected to operate relatively freely and openly, so that firms experience no substantial barriers to the international transfer of finished or semifinished goods. This stands in contrast to the product cycle model, in which barriers to entry motivate firms to invest abroad. Using a coarse categorization of nations into center (developed countries highly integrated into the world economy), periphery (countries poorly integrated into the world economy), and semi-periphery (neither central nor peripheral, with intermediate levels of wages and living standards), Fröbel, Heinrichs, and Kreye argue that firms will exploit differences in wages and labor skills to generate higher profits. In particular, they suggest that large firms will fragment production processes into stages such that each portion can be located in that area in which the costs to the firm are minimized (see Table 5–2).

As a result, they expect that activities that will be retained within central economies are those requiring high education levels and labor skills, such as high technology production (particularly in its early stages) and research and development. Peripheral economies, because of the combination of low wages and low skill levels, are expected to be allocated activities that are labor intensive portions of production. Following processing, reexport is expected, either because the output goods will tend to be semi-processed or because the prevailing local wage is not sufficiently high to develop a local market. Activities requiring lower skill levels than those located in central economies, but higher than those in peripheral economies are expected to be located in the semi-periphery. The outcomes of implementation of this locational logic by large firms, under the assumption that significant portions of the global economy are integrated within these firms, are increasing unemployment in central economies as production activities are relocated to lower wage locations accompanied by increased processing activity with minimal economic or social benefit in less developed countries as the activities transferred will be integrated into the firm's internal activities and thus tend to develop few or no linkages with the local economy. In addition, because the fragmented elements of the production process are executed with well-established production technologies, they tend to be relatively encapsulated activities, and will thus result in few beneficial spinoff activities.

Table 5–2. International Distribution of Activities in NIDL

Location	*Dominant Production*	*Dominant Job Types*
Center	R&D; "hightech"; non-standard	Technical/high-skill requiring high education levels
Periphery	Export-oriented and labor-intensive; often partial processing rather than complete production	Low skill/training requirements; low levels of eduation needed
Semi-Periphery	Routine production requiring higher skills than in periphery	Intermediate skills/training requirements; intermediate education levels needed

Source: After Fröbel, Heinrichs, and Kreye, 1980.

Critics of NIDL have pointed out that while some relocation of activities to peripheral countries has occurred, measures of foreign direct investment still indicate that the bulk of investment in manufacturing is made within the central economies, and thus tends to be linked to either market access or relatively high labor skill requirements, rather than to a strategy based on labor de-skilling and cost minimization (Schoenberger, 1988). In addition, in discussions of the shift to more flexible production, it has been argued that the higher skilled R&D functions within firms should be viewed as complementary to routine production, and that physically separating these functions can have negative effects on a firm's ability to compete in national or international markets (see Piore and Sabel, 1984; Sabel, 1989).

Since the mid-1970s, models of industry and firm development have been replaced by discussions of types of industrial organization. These discussions include analyses of new dominant forms of firm organization, interactions between firms that may herald the replacement of more distant exchange relations with more interactive cooperation, and the extent to which new production technologies, methods of inventory control, and ways of organizing work flow will change internal organization, relations among firms, and the spatial organization of the production process. In particular, as firms in established industries in developed nations have watched the rise of new sectors with distinctly different industrial organization patterns and supplier relations, and the rise of foreign competitors using different work flow, inventory, and supplier relations, interest has shifted from identifying patterns which were successful in the past to identifying the new leading patterns and the characteristics that are the keys to their industrial competitiveness.

While the older leading views of firm development focused on the development of increasingly hierarchical administrative structures and the increasingly large geographical extent of production and marketing, newer views emphasize different strategies, structures, and geographies. New "leading models" of industrial organization come from three sources: firms that have become the new leaders of established industries, firms within industrial branches that have developed recently, and firms within newly developed or particularly resilient sub-national regions. For established industries, the practices of new lead firms have been contrasted with those of less dynamic firms in order to identify key elements of competitive success. For example, numerous aspects of Japanese automobile assemblers, particularly Toyota and Honda, have been contrasted with those of U.S. and European assembly firms (Mair, Florida, and Kenney, 1988; Sadler, 1994). When, as in the automobile industry, the firms from a single nation appear to be the new lead firms, larger sectoral structures, practices, and institutions have been included in the comparative evaluation. In new industries, such as microelectronics, the industrial organization of the branch as well as the structure and strategy of the largest and most dynamic firms have been the focus (for example, Garud and Kumaraswamy, 1993). At the regional level, the organization of regional industries, including firm interactions and labor market organization, has been seen as key to economic stability and possible self-sustaining development (see Scott, 1988, 1990, 1991; and Scott and Angel, 1987 for examples; and Hirst and Zeitlin, 1991; Amin and Malmberg, 1992; and Lipietz, 1993 for critical reviews).

These new models of organization in firms and industries have led to debates over the extent to which new organizational forms represent breaks with the past and foreshadow the future contours of industrial organization in developed countries. Some, including Piore and Sabel (1984), Sabel (1989), and Scott (1988) have argued that new industrial spaces and resilient regions in which industrial complexes are concentrated have, by being centers of competitive industries and firms, shown such advantages during a period of economic uncertainty and instability that they clearly represent the best alternative for future development. Others, such as Lovering (1990) and Gertler (1988), on the other hand, have pointed out that many of the characteristics linked to these regional complexes are not new, and that alternative industrial formations may also be competitive.

A result of these debates has been to broaden the discussion beyond the obvious industrial leaders and new growth regions to include adjustment processes in established industries and less central or dominant production regions. These have shown that recent changes have involved incorporating some new elements into a path of development that exhibits more continuity than abrupt structural change (see Dunning, 1993, for an overview). In particular, while some large firms, such as IBM, have undergone abrupt organizational restructuring and reorientation of production and sales, most have adopted strategies of incremental change. For large firms, adaptation strategies clearly indicate that increased competition has led to renewed emphasis on cost reduction for existing products, with an emphasis on the adoption of the best industrial practices to ensure output quality. Because markets for current output are viewed as changing in demand levels and subject to shorter product cycles, cost reduction has been complemented by attempts to increase flexibility in the volume and variety of output, as well as raise the level of innovation by developing both new variants of existing and entirely new products. To achieve the twin goals of increased flexibility and innovativeness, large firms are seeking cooperative partners for research and development, production, and marketing more often than in the past (see Doz,

1992). To some extent, the development of cooperative activities reflects the financial stress firms endure during periods of increased competition. Cooperative activities have been less expensive alternatives than the acquisition of businesses with complementary assets, but have played the same role in firms' restructuring strategies.

GEOGRAPHIES OF PRODUCTION IN LARGE FIRMS

Recent trends in the organization and operation of large firms has shown that, while firms tend to face similar competitive conditions, there have been a diversity of geographical outcomes. In many cases, the resulting spatial patterns of production are strongly conditioned by production and exchange characteristics and result in particular spatial patterns at the regional, or micro scale, while at the global, or macro scale, the firm's market strategy plays a significant role.

With respect to the internal organization of firms, one component of change appears to hold consistently: the internal organization of firms in both established and new industries allows for more rapid responses to change and for a quicker flow of information internally. As Eggertsson (1990) has pointed out, this is in part because of the general shift during the past 25 years in the world economy to the production of goods which involve the horizontal flow of multiple components, rather than the vertical processing of bulk commodities. Because large firms in industries with simpler, vertical production flows have tended to shift forward into more complicated processing stages—for example bulk steel producers have incorporated downstream processing stages of capital goods production and mechanical engineering, or have diversified into more dynamic product line—few firms remain simple bulk processors. The result of these general secular changes is that while the multidivisional corporation that Chandler evaluated may have directed mass processing of materials well, the more intricate and complicated exchanges associated with the industries that have predominated since the 1970s and 1980s require more distributed decision making and control. These changes have given rise to the "network" firm, in which information flows not only vertically within the operating units of the firm, but also horizontally between units (see Garud and Kumaraswamy, 1993, for an example of an "open" network, and Bartlett and Ghoshal, 1993, for an internal, or closed, network).

In the following subsections, several of the geographical patterns associated with large firms will be discussed. First, the micro-scale phenomena of the regional reconcentration of production will be presented. Second, this is set in the context of global production patterns. Third, the process of "hollowing out" through a shift toward distributed global production is outlined.

Regional Reagglomeration

Since the late 1970s, the adoption of just-in-time inventory control, along with changes in production technologies that have reduced the minimum efficient scale of production in many assembly industries, has resulted in shifts in the spatial organization of production in some industries from a dispersed network of suppliers to a pattern with a stronger regional concentration of linked activities. In some cases, industries which were expected to adopt the NIDL approach by distributing stages of production internationally have been among those that have demonstrated strong tendencies to develop or redevelop regional centers of industrial activity. This is particularly true for the automobile industry which, especially for North American producers, had earlier seemed to be adopting international sourcing and subassembly strategies in order to reduce costs. Now, in contrast to geographically distributed patterns, "hub-and-spoke" patterns (see Figure 5–1), in which component suppliers are located within easy reach, usually one to two hours, of the assembly plant they serve, appear to have become the norm. Supplier plants may belong to the same firm of the assembly plant (integrated) or they may be separate legal entities. The advantages that these plants offer the final assembler are lower cost operating expenses associated with the reduction in component inventory and the ability to detect and correct problems in supply components before the investment in high stock volumes is made. Because of the short distances traveled between the supplier and the plant, there is little risk to the logistics of supply flow.

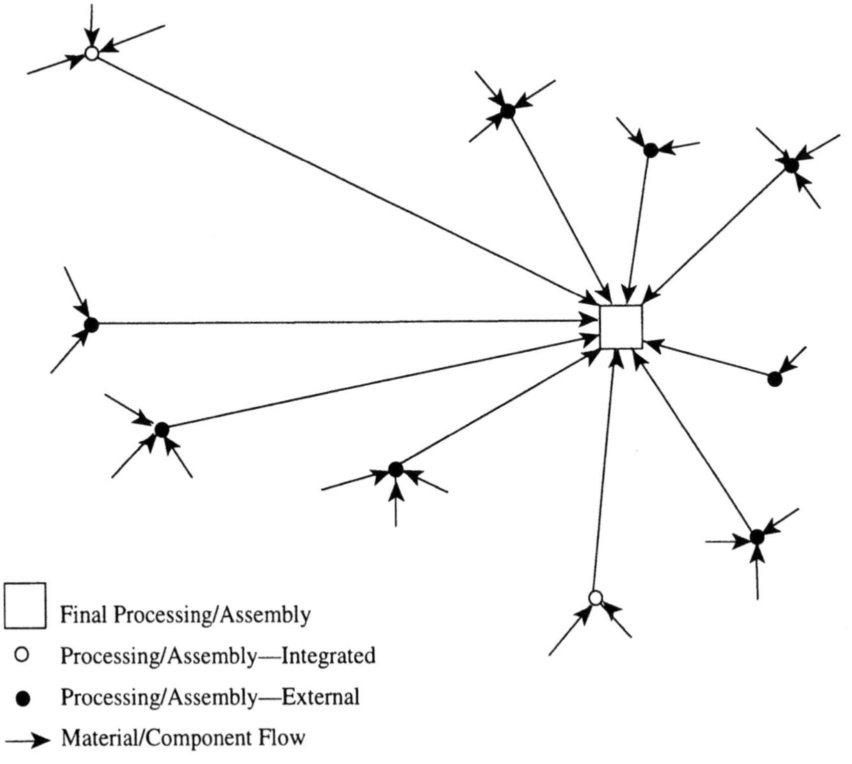

Figure 5–1. Hub-and-Spoke Configuration

To be a more effective strategy than global sourcing of components, two conditions must hold. First, the benefits of regionalized supply must outweigh the lower costs of a more dispersed network. That is, in the localized pattern, the interaction between the final assembler and the suppliers must, through reduction in inventory costs and/or improvements in quality, outweigh the combination of lower absolute labor costs and lower bulk prices plus higher freight inventory costs. In addition, the advantages in the concentrated pattern of more flexible delivery schedules that can be orchestrated in a short time horizon and the potentially higher risks to production disruption, including labor unrest, when activities are concentrated in a single area must be weighed. Second, it must be possible to induce suppliers to locate in the region surrounding the final assembly plant. Clearly, this is not a consideration for integrated supplier plants, which are directly controlled by the firm, but may become an issue for external suppliers if the development of a new supply location requires substantial capital investment. For some supplier firms, developing a new facility may require contractual commitment by the assembler. Should problems subsequently arise, such a commitment may impair the ability of the assembler to shift to another supply source. For automobile assemblers in North America during the 1980s and 1990s, inducing suppliers to develop warehousing or supply plants does not appear to have been problematic. This may be for two general reasons. First, technological changes may have reduced the necessary scale of production so that plant size and cost were reduced. Second, due to the generally poor conditions faced by domestic automotive component suppliers in the 1980s and 1990, the development of new facilities may have appeared a relatively attractive way to increase receipts, despite the cost of investment. Although some assemblers have chosen traditional auto plant locations such as Detroit, within the U.S. the development of new plants, particularly by Japanese assembly firms, has led to a shift in production facilities to more rural areas generally near but extending from older concentrations of production (Mair, Florida, and Kenney, 1988).

While the implementation of the hub-and-spoke pattern in the U.S. appears particularly strong, it has not been replicated in the investments by Japanese assembly firms within western Europe. There, Japanese investment has occurred later and at a slower pace than in North America. In addition, there has been substantial political pressure to produce automobiles with high levels of components supplied by European producers, so that assemblers have sought to develop relations with existing indigenous suppliers relatively early (Sadler, 1994). The result has been the development of a supply system that is much less regionally concentrated than in North America. Sadler (1994) also notes that, in addition to political pressure, both the extent and quality of existing European suppliers influenced the strategy of using existing suppliers rather than developing a new supply network.

Two Models of International Production

The development of firms that market products in multiple national markets has given rise to two basic models of production (Figure 5–2). The first, multidomestic production, consists of a distributed production system in which the firm serves each major market from a separate supply base. In many respects, this geographical arrangement of production corresponds to that described in the mature phase of the product cycle, in which the firm has extended its production network to serve the major markets developed initially through exports. The second model, global production, corresponds to the production arrangement described in NIDL, in which the production process is geographically fragmented so that subassembly and processing stages are located where they can be carried out most efficiently. The most important aspect of the production process in determining which of these is selected by a firm or dominates an industry appears to be the scale of minimum efficient production in relation to the size of the markets served (Dunning, 1993). In addition, the multidomestic model requires that the necessary input materials must be generally available or relatively inexpensive to transport to the market to be served.

Types of products typically associated with production for a domestic market include those, such as food processing, in which local input supplies and local markets coincide. International producers may thus develop either by establishing new production facilities or by acquiring existing national firms. In the case of Japanese automobile assembly firms, for example, expansion into local markets has been through the development of new production facilities, integrated with existing suppliers in the case of western Europe but generally with new supplier facilities in the case of the U.S. (see above). Because the scale of production in western Europe is generally larger than the demand in any single nation, the final assembly plants serve a multi-nation region rather than a single domestic market.

The global model of production, on the other hand, is generally expected to dominate in industries in which the advantages of sourcing input materials outweighs the disadvantages of shipment costs and time delays. For industries that serve mass markets with relatively stable types of output, the advantages may lie with a lower cost, geographically extensive production network.

Industries typically associated with this model of production include textiles and apparel and electronics. The textile and apparel industry has become globalized, with international networks of production serving extensive mass markets (Fröbel, Heinrichs, and Kreye, 1980; Schatz and Wolter, 1987). The production of standard cloth is no longer dominated by industrialized nations. However, the production of specialty fabrics with capital intensive equipment has been retained in core countries. In some cases, cloth is produced and laser cut in highly automated facilities in developed countries, then exported to low labor cost locations such as Haiti or Mauritius for apparel construction, then distributed to multiple international markets. More specialized manufacture of niche products with short production runs, however, has tended to remain in domestic markets (Schatz and Wolter, 1987). Scott and Angel (1987) have noted a similar "dual" production structure in microelectronics among U.S. firms, with the production of more standardized devices organized in spatially distributed networks, while more innovative, specialty devices are produced within regionally concentrated production complexes of disintegrated producers.

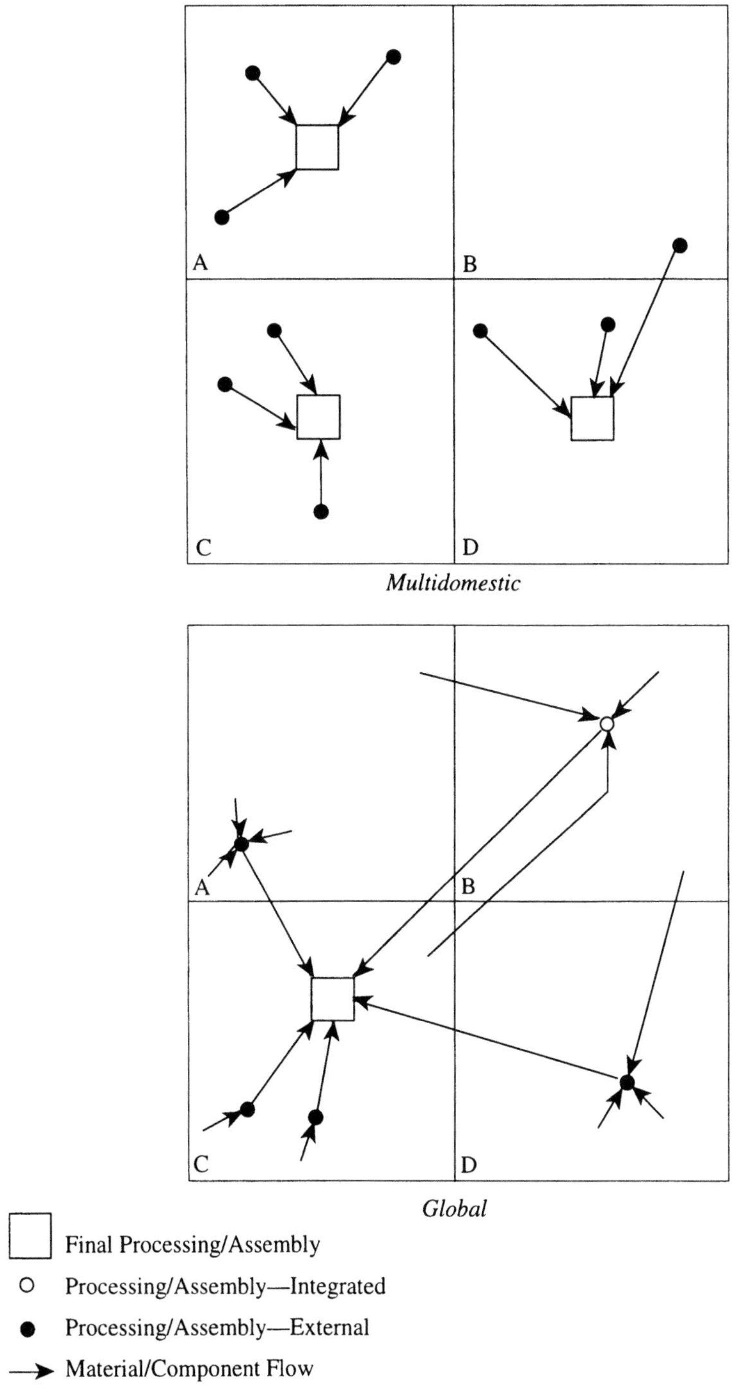

Figure 5–2. Prototypes of International Production

Global Sourcing and "Hollow" Corporations

The implementation of global production, by U.S. firms in particular, has led to concerns of overly rapid de-industrialization in the U.S. compared to other developed nations, loss of manufacturing employment, and possible loss of competitiveness by U.S. producers as control over the production process is lost. Global sourcing is described as leading to "hollowness" in two ways. First, U.S. operations of a firm can become hollow as portions of the production process are relocated to foreign countries. Global sourcing results in U.S. operations comprising administrative and marketing personnel but few production activities. In this case, the firm retains control over the relocated activities. Second, the firm may become hollow in a more general sense when a firm substitutes external foreign suppliers in place of its internal, domestic sources. In this case, the firm may lose control over and information about the externalized activity. Figure 5–3 presents an example of progressive hollowing of a home country producer, as it shifts from being a domestic producer, using few imported materials, to using foreign subassembly (stage 2), shifting its final assembly plant abroad (stage 3), and sourcing additional inputs externally (stage 4).

The evaluation of the sourcing strategies of large, multinational manufacturing firms indicates that the process of "hollowing out" has been stronger for U.S. than for European or Japanese firms (Kotabe, 1992). In part, this reflects the earlier development of international production networks by U.S. firms and the competitive disadvantages the U.S. as a production location faced as the dollar appreciated. In addition, however, this reflects a somewhat different approach to global production, in which U.S. firms have tended to shift component sourcing for foreign assembly while European and Japanese retained more home country production (Kotabe, 1992, Chapters 2 and 7).

While shifting to a global production pattern is expected to reduce a firm's costs by permitting firms to incorporate resources more efficiently, global sourcing with external suppliers is viewed as containing at least potential threats to the firm's longer term competitive position. If global sourcing results in firms losing control over and information about a production process, they may miss the opportunity to innovate and improve both process and product.

In addition, if the component is a significant element in the product, as the supplier becomes familiar with the production process, a skillful competitor may develop.

On the other hand, if the production of key components is relocated globally, but not externalized, and firm managers are able to coordinate production and integrate learning from the foreign-based production site with product development based elsewhere, the hollowing process may present no significant long term threat. The coordination of R&D with production may not, however, easily occur when these are physically separated. If effective coordination requires physical proximity, this may lead either to a dispersion of R&D activities in the firm or to the relocation of production closer to centralized R&D facilities (Kotabe, 1992, p. 179; Schoenberger, 1987).

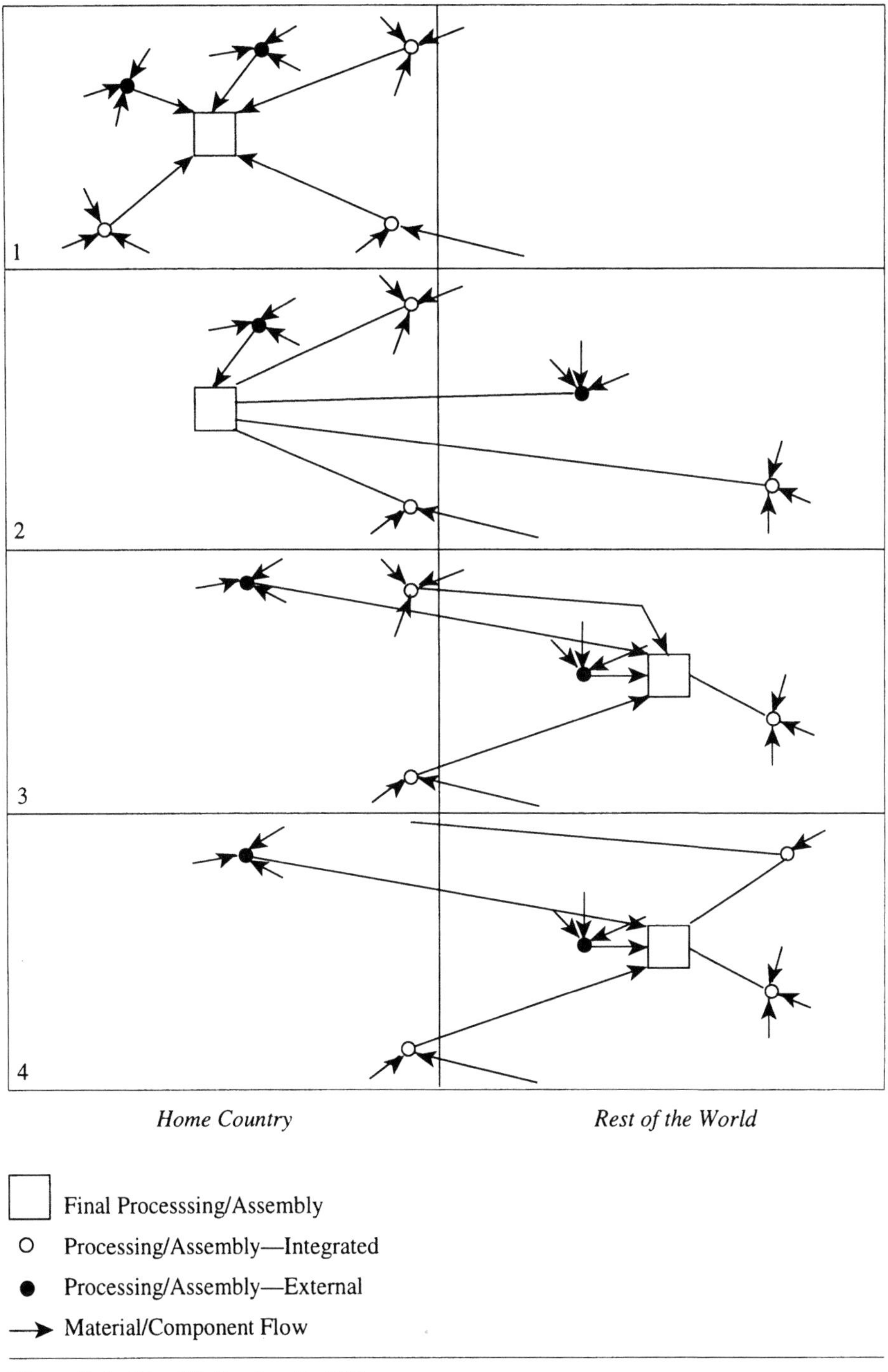

Figure 5–3. The "Hollowing Out" Process

CONCLUSION

The patterns of production in large firms that have emerged since the early 1970s represent both continuity and change. The earlier patterns of international production, distilled into models such as the product cycle and NIDL, represented "pure" forms of a particular locational logic. The development of the world economy and of production technology since the 1970s, however, has introduced spatial patterns of production with a broader variety of forms. In some cases, the result has been a reconcentration of production in regions, while in others the development of spatially extensive supply and production networks has continued.

REFERENCES

Amin, A. and Malmberg, A. (1992). "Competing Structural and Institutional Influences on the Geography of Production in Europe," *Environment and Planning A*, 24, pp. 401–16.

Bartlet, C. and Ghoshal, S. (1993). "Beyond the M-form: Toward a Managerial Theory of the Firm," *Strategic Management Journal*, 14, pp. 23–46.

Bell, D. (1973). *The Coming of Post-Industrial Society*. New York: Basic Books.

Boyer, R. (1988). *The Search for Labour Market Flexibility: The European Economies in Transition*. Oxford: Clarendon.

_____. (1990). *The Regulation School: A Critical Introduction*. New York: Columbia

Chandler, A. (1962). *Strategy and Structure*. Cambridge: MIT.

_____. (1977). *The Visible Hand: The Managerial Revolution in American Business*. Cambridge: Harvard University.

_____. (1990). *Scale and Scope: The Dynamics of Industrial Capitalism*. Cambridge: Harvard University.

Doz, Y. (1992). "The Role of Partnerships and Alliances in the European Industrial Restructuring." In K. Cool, D. Neven, I. Walter (Eds.), *European Industrial Restructuring in the 1990s*. New York: New York, pp. 294–327.

Dunning, J. (1979). "Explaining Changing Patterns of International Production: In Defence of the Eclectic Theory," *Oxford Bulletin of Economics and Statistics* 41, pp. 269–95.

_____. (1993). *Multinational Enterprises and the Global Economy*. Wokingham: Addison Wesley.

Eggertsson, T. (1990). *Economic Behavior and Institutions*. New York: Cambridge University.

Foss, N. (1993). "Theories of the Firm: Contractual and Competence Perspectives," *Journal of Evolutionary Economics* 3, pp. 127–44.

Fröbel, F.; Heinrichs, J.; and Kreye, O. (1980). *The New International Division of Labor*. Cambridge: Cambridge University.

Garud, R. and Kumaraswamy, A. (1993). "Changing Competitive Dynamics in Network Industries: An Exploration of Sun Microsystems Open Systems Strategy," *Strategic Management Journal* 14, pp. 351–69.

Gertler, M. (1988). "The Limits to Flexibility: Comments on the Post-Fordist Vision of Production and its Geography," *Transactions of the Institute of British Geographers, N.S.* 13, pp. 419–32.

Hirst, P. and Zeitlin, J. (1991). "Flexible Specialization Versus Post-Fordism: Theory, Evidence and Policy Implications," *Economy and Society* 20, pp. 1–56.

Jacquemin, A. and de Jong, H. (1977). *European Industrial Organization.* New York: John Wiley.

Kotabe, M. (1992). *Global Sourcing Strategy.* New York: Quorum.

Lipietz, A. (1993). "The Local and the Global: Regional Individuality or Interregionalism?" *Transactions of the Institute of British Geographers N.S.* 18, pp. 8–18.

Lovering, J. (1990). "Fordism's Unknown Successor: A Comment of Scott's Theory of Flexible Accumulation and the Re-emergence of Regional Economies," *International Journal of Urban and Regional Research* 14, pp. 159–74.

Mair, A.; Florida. R.; and Kenney, M. (1988). "The New Geography of Automobile Production: Japanese Transplants in North America," *Economic Geography* 64 , pp. 352–71.

Noel, A. (1987). "Accumulation, Regulation, and Social Change: An Essay on French Political Cconomy," *International Organization* 41, pp. 303–33.

Piore, M. and Sabel, C. (1984). *The Second Industrial Divide: Possibilities for Prosperity.* New York: Basic Books.

Rostow, W. (1960). *The Stages of Economic Growth.* Cambridge: Cambridge University.

Sabel, C. (1989). "Flexible Specialization and the Re-emergence of Regional Economies." In P. Hirst and J. Zeitlin (Eds.), *Reversing Industrial Decline? Industrial Structure and Policy in Britain and her Competitors.* Oxford: Berg, pp. 17–70.

Sadler, D. (1994). "The Geographies of Just-in-Time: Japanese Investment and the Automotive Components Industry in Western Europe," *Economic Geography* 70, pp. 41–59.

Schatz, K. and Wolter, F. (1987). *Structural Adjustment in the Federal Republic of Germany.* Geneva: International Labour Office.

Scherer, F. and Ross, D. (1990). *Industrial Market Structure and Economic Performance* (3rd ed.). Boston: Houghton Mifflin.

Schettkat, R. (1992). *The Labor Market Dynamics of Economic Restructuring.* New York: Praeger.

Schoenberger, E. (1987). "Technological and Organizational Change in Automobile Production: Spatial Implications," *Regional Studies* 21, pp. 199–214.

_____ . (1988). "Multinational Corporations and the New International Division of Labor: A Critical Appraisal," *International Regional Science Review* 11, pp. 105–19.

Scott, A. (1988). "Flexible Production Systems and Regional Development: The Rise of New Industrial Spaces in North America and Western Europe," *International Journal of Urban and Regional Research* 12, pp. 171–86.

_____. (1990). "The Technopoles of Southern California," *Environment and Planning A* 22, pp. 1575–605.

_____. (1991). "Electronics Assembly Subcontracting in Southern California: Production, Processes, Employment, and Location," *Growth and Change* 22, pp. 22–35.

Scott, A. and Angel, D. (1987). "The U.S. Semiconductor Industry: A Locational Analysis," *Environment and Planning A* 19, pp. 875–912.

Vernon, R. (1966). "International Investment and International Trade in the Product Cycle," *Quarterly Journal of Economics* 80, pp. 190–207.

_____. (1979). "The Product Cycle Hypothesis in a New International Environment," *Oxford Bulletin of Economics and Statistics* 41, pp. 25–67.

6

SERVICES AND FLEXIBLE PRODUCTION: CONTINGENT BUT MUTUALLY REINFORCING SYSTEMS

Barney Warf
Department of Geography, Florida State University, Tallahassee, FL 32306-2050 USA

INTRODUCTION

Analyses of economic change in the late twentieth century frequently point to two parallel sets of phenomena, the emergence of so-called "flexible production" systems and the explosive growth of services. Both of these have radically reshaped what is produced in various parts of the global economy, how it is produced, who produces it, and where production is located. While the growth of the service economy occurred for reasons only partly related to the emergence of flexible production systems, services would not be as prominent as they are today without the emergence of flexible production. In short, services and flexible production are contingent, but mutually reinforcing systems.

The relations between services and flexible production systems are complicated by the casual, often incorrect use of both terms (reminiscent of the days when both were classified as "high tech"). Both terms are ambiguous, encompassing a variety of products, processes, and occupations. Further, whatever definition is employed, there are many obvious similarities between the two: services often exhibit characteristics of flexible production, including the presence of many small firms, part-time employment, technological sophistication, and localized networks of subcontractors. However, services generally suffer from a poor public image (e.g., low-paying hamburger flippers), while flexible production systems, at least in the popular image of "high-tech" industries, enjoy a great glossy patina as a panacea for unemployment, a generator of skilled, environmentally clean jobs, and as an innovative sector restoring U.S. competitiveness internationally.

Difficulties in studying services are compounded by a long-standing bias against the sector. Despite the vast size of services (70% of the labor force in most industrialized nations), works addressing this issue pale compared to the mountain of literature on the automobile or steel industries, or for that matter, flexible production. This obsessive focus on manufacturing reflects a long-standing bias against services, which have been dismissed as "unproductive" since the days of Adam Smith, who regarded them as "residual" activity in the face of agriculture and manufacturing (Stanback, 1979). Neither has contemporary social theory been particularly comfortable with the service sector. Unlike manufacturing, with its Dickensian images of a sweaty, male, collectivized proletariat, services do not lend themselves easily to Marxist analysis: how does the labor theory of value measure the exploitation of a secretary? Many theorists have ignored the issue altogether. Where Marxists have turned to services, their opinion has been largely dismissive (e.g., Walker, 1985). Thrift (1987, p. 204) notes that "there has been a general tendency in Marxist and Marxist-related theory, from Marx onwards, to downgrade the importance of commercial capital within capitalism by seeing it as a 'parasitic,' or even 'relic' form to be relegated to the sidelines of capitalist production." Harvey (1989), for example, relegates finance to trade in "fictitious capital," i.e., "capital that has a nominal money value and paper existence, but which at a given moment in time has no backing in terms of real productive activity or physical assets as collateral. Fictitious capital is converted into real capital to the degree that investments are made that lead to an appropriate increase in useful assets (e.g., plant and machinery that can be profitably deployed) or commodities (goods and services which can be profitably sold)" (p. 182).

Harvey equates productive capital with commodity production (the "primary circuit" of capital accumulation), dismissing finance, and by implication, all services, as "unproductive." Here Harvey falls in line with a long tradition of export-base studies that equate services with local consumption, when in fact services are frequently exported from cities and regions (e.g., Beyers and Alvine, 1985; Goe, 1990; Stabler and Howe, 1988).

In the face of such definitional and conceptual difficulties, this chapter argues for the autonomy and significance of services. It is particularly concerned with the multiple ways in which the service economy and flexible production systems are shot through with each other. Contrary to teleological views that hold advanced economies progress through stages from manufacturing to services, the perspective advanced here holds there is no necessary linkage between them, i.e., flexibility was not a necessary condition for the growth of services, nor do services inevitably replace regional or national manufacturing bases. Rather, this chapter holds that the relations between these two domains are, in realist terms, contingent, not necessary, in character (Sayer,

1984). Thus, services and flexible production systems arose for distinct sets of reasons, each with its own source of growth. As there is an abundant literature on flexible production (e.g., Scott and Storper, 1987; Scott, 1988; Amin and Robins, 1990), this chapter shall focus on services. However, it is equally evident that each domain of activity has influenced the other, the subject of scrutiny here.

The chapter opens with a straightforward investigation into the definition and composition of services. Next, it outlines the major causes of growth of services employment and output nationally and internationally, stressing that these bear few linkages to the emergence of flexible production. Third, it turns to the impacts of flexible production upon the generation of services and the question of services externalization. Fourth, it points to the impacts of the digital revolution on the geography of producer services, particularly the role of telecommunications in changing the global geography of finance and clerical employment.

WHAT ARE SERVICES?

"Services" encompass an enormous diversity of occupations and industries (Daniels, 1991), ranging from professors to plumbers to prostitutes. Indeed, so great is the variation among firms and occupations within services that the term threatens to lose any coherence whatsoever. Some observers argue that services include all information processing activities, including clerical activity, executive decisionmaking, telecommunications, and the media (Hall and Preston, 1988). While many service jobs do involve the collection, processing, and transmission of large quantities of data, clearly others do not. The trash collector, restaurant chef, security guard, and janitor all work in the service sector, but the degree to which these activities center around "information processing" is minimal.

Others emphasize that services involve the production of intangible outputs, and thus stand in contrast to manufacturing, the product of which can be "dropped on one's foot." What, for example, is the output of a lawyer? a teacher? a doctor? It is impossible to measure these outputs accurately and quantitatively, yet they are real nonetheless. The U.S. government estimates services output using revenues as a proxy. Yet revenues are determined by output and prices, so this measure is distorted by changes in the relative prices of service outputs. To complicate matters, many services generate both tangible and intangible outputs. Consider a fast-food franchise: the output is assuredly tangible, yet it is considered a service; the same is true for a computer software firm, in which the output is stored on disks. The fact that it is difficult, if not impossible, to measure output in services has enormous implications for their analysis. For example, some critics of the service sector argue that the slowdown in U.S. productivity growth in the late twentieth century reflected the growth of services. Yet if output in services cannot be adequately measured, how can one argue that output per employee in services is high or low, rising or falling? If wages reflect productivity increases, what is the relation between services output and salaries?

If outputs do not differentiate services from manufacturing, then what about inputs? All forms of manufacturing as well as services involve inputs of *both* goods and services: an automobile producer must purchase legal services, advertising, banking, and public relations inputs, just as airlines or securities firms must purchase hardware and equipment to do what they do (Walker, 1985; Coffey and Polèse, 1989). Clearly the lines between services and manufacturing are blurry and no simple definition will suffice. Despite these definitional problems, a broad consensus exists as to the major components of the service sector:

a. *Finance, insurance, and real estate*: commercial and investment banking, insurance of all types, and the commercial and residential real estate industry.

b. *Business services*: legal services, advertising, engineering and architecture, public relations, accounting, research and development, and consulting. Many producer services also serve final demand, such as attorneys that cater to both commercial clients and individuals.

c. *Transportation and communications*: the electronic media, trucking, shipping, railroads, airlines, and local transportation (taxis, buses, etc.).

d. *Wholesale and retail trade*: intermediaries between producers and consumers. Closely affiliated are eating and drinking establishments, personal services, and repair and maintenance services.

e. *Entertainment, hotels and motels*: elements of tourism, the world's largest industry.

f. *Government at the national, state, and local levels*: public servants, the armed forces, and all those involved in the provision of public services (e.g., public education, health care, police, fire departments, etc). While much has been written about the state, its forms and functions, the public sector has only recently been viewed as part of the service economy.

g. *Nonprofit agencies:* charities, churches, museums, and private, nonprofit health care agencies, many of which play influential roles in local economies.

Every definition of "services" is slippery, however. For example, does "services" refer to a set of industries or occupations? (The U.S. government sidesteps this problem by adopting industries.) Many workers in manufacturing are in fact service sector workers, including personnel in headquarters and clerical functions and research. Is the secretary who works for an automobile company part of manufacturing, while the secretary who works for a bank part of the service sector? The use of industrial versus occupational definitions is particularly critical given the increase in "nondirect" production workers within manufacturing. Clearly, different definitions of "services" have significantly different implications for assessments of the size and composition of the service sector.

FORCES DRIVING THE GROWTH OF SERVICES

The rapid growth of services, and their relations to flexible production, must be viewed within the context of the enormous series of changes undergone by late twentieth century capitalism (Harvey, 1989). The signs of this transformation include: the collapse of the Bretton-Woods agreement in 1971 and the subsequent shift to floating exchange rates; the oil crises of 1974 and 1979 and the resulting recession and stagflation in the West; the explosive growth of Third World debt; the steady deterioration in the competitive position of industrial nations, particularly the U.S. and the U.K., and the concomitant rise of Japan, Germany, and the Newly Industrializing Nations, particularly in East Asia; the emergence of flexible production technologies (e.g. just-in-time inventory systems) and the computerization of the workplace; the steady growth of multinational corporations and their ability to shift vast resources across national boundaries; the global wave of deregulation and privatization that lay at the heart of Thatcherite and Reaganite post-Keynesian policy; and finally, the integration of national financial markets through telecommunications systems (Brunn and Leinbach, 1991). Independent of how this series of changes is labeled (compare Graham et al., 1988; Harrison and Bluestone, 1988; Marshall, 1988), what is abundantly clear from these observations is the emergence of a new global division of labor, in which services and flexible production systems play a fundamental role.

Given the emergence of the post-Keynesian state, most obviously in the U.S. and the U.K., it is curious how little attention has been paid to the extent and impacts of deregulation. The lifting of state controls on transportation and communications, for example, has had significant impacts on the profitability, industrial organization, and spatial structure of numerous sectors. In the U.S, the shipping, trucking, and aviation industries underwent sustained deregulation and telecommunications were reorganized following the dissolution of A.T.&T.'s monopoly in 1984. Similarly, banking, insurance, and securities also experienced a relaxation of state controls. In the U.S., these moves included the removal of interstate banking regulations (Holly, 1987), the end of laws governing pension and mutual fund portfolios, the abolition of fixed commissions on stock market transactions, and the current debate over the repeal of the Glass-Steagall Act, which legally separates commercial from investment banking.

Why, then, have services grown so rapidly in the twentieth century? Services employment has increased steadily in the face of low rates of population growth, slowly rising rates of productivity and income, and significant manufacturing job loss. While services and manufacturing are intimately intertwined, it is equally apparent that services exhibit growth and locational dynamics somewhat different from manufacturing (Kirn, 1987), although both constitute commodity production in varying forms. Five reasons for the increase in services employment throughout the world are suggested below. A sixth, externalization, is covered later.

First, rising per capita incomes, particularly in the industrialized world, have contributed markedly to rising services employment (Ginzberg and Vojta, 1981). The demand for many services is income-elastic. That is, increases in real personal income tend to generate proportionately larger increases in demand (in contrast to most manufactured goods). Services with particularly high income-elasticities include entertainment and transportation, which largely explains why tourism is the largest industry in the world today (Tuttle, 1987). In the late 1980s, U.S. households, for the first time in their history, spent slightly more on services (51% of disposable income) than they did on durable and nondurable goods combined. An important reason contributing to this growth is the increasing value of time that accompanies rising incomes (especially with two-income families). As the value of time climbs relative to other commodities, consumers generally will attempt to minimize the time inputs needed for the accomplishment of ordinary tasks. While this phenomenon also explains the demand for washing machines, dishwashers, and automobiles, it is especially important for the growth of services. The explosion of fast-food restaurants, for example, has little to do with the quality of food (or even the price), and much to do with attempts by consumers to avoid cooking at home or shopping. Similarly, the growth of repair services reflects both increasingly sophisticated technologies and a generalized unwillingness to spend limited recreation time doing such chores. Thus, the increasing value of time has led to a progressive externalization of household functions, so that which used to be done in-house becomes a commodity purchased through the market for a profit.

Second, rising levels of demand for health and educational services comprise an important part of the broader growth of the service economy. The provision and consumption of health care has increased steadily, in large part because of the changing demographic composition of the populations of industrialized nations. The most rapidly growing age groups today are the middle-aged and the elderly, precisely those demographic segments that require relatively high per capita levels of medical care. Consequently, medical services as a proportion of GNP have increased steadily throughout Europe, North America, and Japan, often leading to political and economic conflicts about how to contain them. Similarly, a changing labor market and increasing demand for literacy and numeracy at the workplace have driven the increasing demand for educational services at all levels, a process reflected in higher enrollments in universities, which have become prerequisites for entrance to middle class jobs.

Third, the growth of services reflects the rising proportion of nondirect production. Most corporations today devote considerable resources to dealing with a complex marketplace and legal environment, including many specialized clients, complex tax codes, environmental and labor restrictions, international competition, sophisticated financial markets, and complex real estate purchases and sales. To do so, they require administrative bureaucracies to collect and process information and make strategic decisions, clerical workers to assist with mountains of paperwork, sales people, researchers to study market demand and create new products, and legions of advertisers, public relations analysts, accountants, lawyers, and financial experts to assist in an enormously complicated decisionmaking environment. Similarly, the introduction of sophisticated machinery requires maintenance and repair staff, while offices or industrial plants require security and building maintenance staff—all nondirect production workers, and all services.

A fourth reason underpinning the growth of services is the increasing size and role of the public sector and concomitant expansions in government employment. Despite oft-repeated insistences that capitalism is governed by market relations, the state in fact acts as a major actor shaping markets, subsidizing firms, and framing the broader context of economic activity. The government contributes to the growth of services in two ways. First, government employment has increased steadily, especially since the New Deal of the 1930s, because the public demands the services that it provides. Today, the federal government is the largest single employer in the U.S. employing more than 2 million people, and although federal employment grew only slowly in the 1980s, it is dwarfed by local and municipal government employment. Government also contributes to the growth of services indirectly, through a labyrinthine web of laws, rules, restrictions, and regulations, contributing to the growth of tax attorneys, accountants, consultants, and other specialists that assist firms (externally or internally) in negotiating with the legal environment.

A fifth reason for the growth of services is rising levels of service exports within and among nations. A widespread myth exists that services by necessity cater to local demand, i.e., they are nonbasic activities and are thus of secondary importance to manufacturing. This notion reflects the long-standing bias against services noted earlier. However, many cities, regions, and nations derive a substantial portion of their aggregate revenues from the sale of services to clients located elsewhere. Many urban areas export services to clients located elsewhere in the same nation (Beyers and Alvine, 1985; Stabler and Howe, 1988; Goe, 1990). Even declining manufacturing regions export services (Harrington and Lombard, 1989). Services are also traded on a global basis, comprising roughly 20% of international trade (Kakabadse, 1987). Internationally, the U.S. is a net exporter of services (but runs major trade deficits in manufactured goods), which is one reason services employment has expanded domestically. Indeed, it could be said that as the U.S. has lost its comparative advantage in manufacturing, it has gained a new one in financial and business services (Noyelle and Dutka, 1988; Walter, 1988). The data on global services trade are poor, but some estimates are that services comprise roughly one-third of total U.S. exports. These sales overseas take many forms, including tourism, fees and royalties, sales of business services, and profits from bank loans. Unfortunately, the U.S. data include repatriated profits from manufacturing investments overseas as a "service," which in reality comprises returns from nonservice activities. Foreign service exports do not generate jobs and revenues in all places equally. Cities such as New York and London, for example, which are critical to global capital markets, have benefitted the most heavily (Thrift, 1987; Moss, 1987a, 1987b).

FLEXIBILITY AND THE EXTERNALIZATION OF SERVICES

The automation of the office beginning in the late 1970s initiated a steady series of technological innovations springing forth from the micro-electronics revolution that fundamentally reshaped the nature of office work and many non-office types of services. Mainframe computers and batch processing initiated a wave of capital intensification that accelerated in the 1980s with the introduction of micro-computers into the workplace, making services major purchasers of flexible production technologies and far more capital intensive than they used to be (Quinn, Baruch, and

Paquette, 1987). One common stereotype of services is that they are uniformly labor intensive. Wages in service industries typically comprise 70–90% of total costs (compared to 10–50% in manufacturing industries), largely because many service jobs are difficult to automate and involve variations in tasks. As Figure 6–1 indicates, however, the value added per employee today among U.S. firms scarcely differs between service and nonservice firms, and among nonretail services, is considerably higher than in manufacturing. Other effects of the micro-electronics revolution upon services include significant gains in productivity (estimated by observers of clerical production to be as high as 30%) in fields such as data processing, medical care, airlines, trucking, and the post office.

Inescapably, this transformation created changes in the requisite skills and occupations. Many old, unskilled jobs were eliminated (e.g., telephone operators), while larger numbers of skilled jobs were generated (e.g., programmers, computer repair). For example, where it once took 14 bank employees to issue a letter of credit,

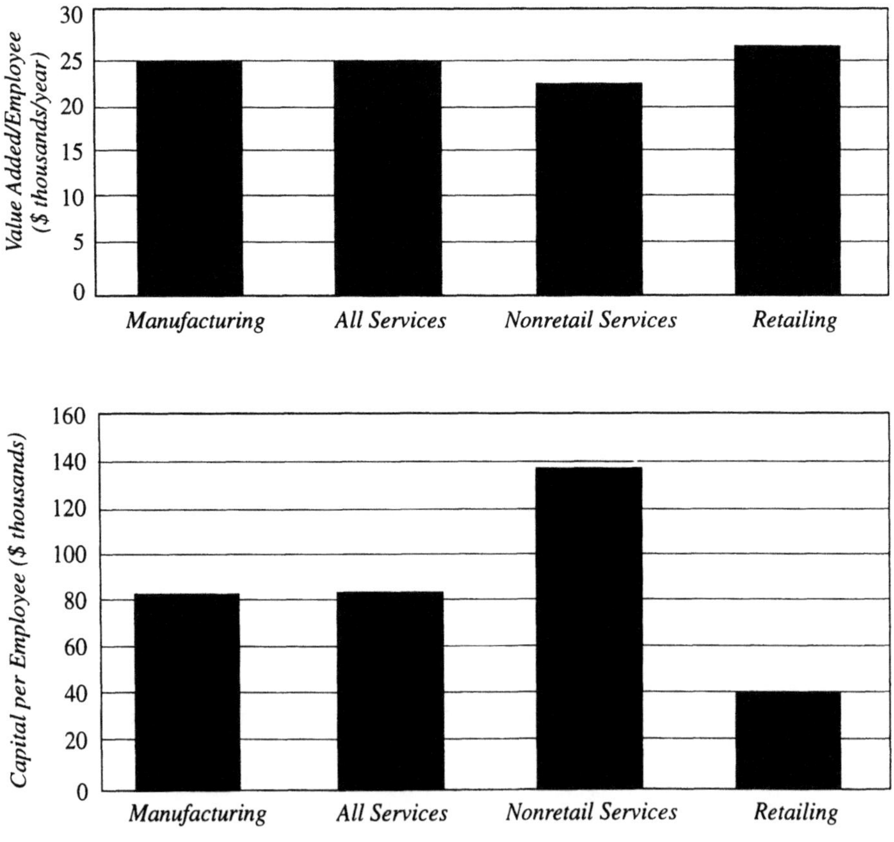

Figure 6–1. Value-Added per Employee in Services and Manufacturing.
Source: Quinn, Baruch, and Paquette, 1987.

one person can now do the same job in minutes. Similarly, bank tellers may now process loan applications (Moss and Dunau, 1986). Computer literacy has become a minimum requirement for employment in services, leading to a net up-skilling in the demand for jobs (in contrast to Marxist visions of widespread de-skilling). Some professional positions, however, saw declines in the skills demanded of them, as when analysts or managers do their own typing on word processing systems, leading to a professionalization of clerical work and a clericalization of professional work (Baran, 1985).

Services also exhibit forward linkages into flexible production systems (Wood, 1991; Marshall and Wood, 1992). Producer service firms primarily offer expertise, reflecting the growing technological and administrative complexity of the workplace. A common argument is that services have grown in part because of the widespread wave of vertical disintegration of production that has occurred globally in the twentieth century, leading to a steady contracting out of many functions from manufacturing firms (Scott, 1983, 1988; Wood, 1991). Vertical disintegration is a persistent theme in the literature on flexible production. Coffey and Bailly (1991) and Goe (1991) suggest the externalization of services occurs through several interrelated processes, including:

a. *product innovation* (transformation of what is produced): e.g., shortening of product life cycles, technological change, emergence of stylized products;

b. *process innovations* (transformation of how things are produced): e.g., process design and engineering;

c. *an increasingly complex financial environment*: growth in activity centered around loans, takeovers, leveraged buy-outs, and foreign exchange;

d. *proliferation of internal management tasks*: strategic planning and coordination that impede the flow of information through the management hierarchy.

They argue further that externalization is most likely when:

a. the firm faces severe in-house technical limitations, i.e., expertise;

b. the firm is an independently owned entity, not a branch plant;

c. the firm is sophisticated relative to its competitors (e.g., technologically, or operates in foreign markets);

d. service inputs are diverse, shifting and non-standardized (e.g., installation, repair, or consulting);

e. the firms face high social insurance costs (e.g., benefits).

Storper and Walker (1989) suggest firms externalize services when they achieve increasing returns from the social division of labor, particularly when they face rising uncertainty, rapid change in products or technology, when the labor process resists easy automation, or when the optimal scales of operation of production processes are markedly different. Thus, from the transactions costs perspective that permeates the literature on flexible production, externalization allows for external economies of scale to replace internal economies of scope.

By externalizing, firms substitute variable costs for fixed ones, and spread the risks of production over their subcontractors, a particularly vital role during peak periods of demand and given mounting levels of uncertainty. Marshall (1982) found that among producer services that firms are likely to externalize, insurance ranks as the most common, followed by accounting, advertising, research, management consulting, advertising and public relations, engineering/architecture, market research, management/professional recruitment, computer installation/repair, commercial law, real estate, and temporary office help. Such services are ones that are not efficient or cost-effective to duplicate in-house, are only needed once or infrequently, and on an unpredictable basis (e.g., trouble-shooting). Typically such services augment or substitute for the internal resources of firms; when the demand is regularized and predictable, they are likely to be internalized within the firm. Further, the degree of externalization depends upon whether the firm is a single establishment or a branch plant, as the latter has the lowest degree of local externalization.

As a result of externalization, many functions formerly performed in-house by manufacturing firms are subcontracted to small suppliers, some of which are in the service sector. Subcontracting and externalization are often studied in the context of manufacturing firms, but have also become increasingly common among large service firms, which often face similar competitive pressures to reduce costs (Storper and Christopherson, 1987).

Statistically, externalization has inflated the growth of services due to the limited languages employed to measure economic activity. For example, a steel company may layoff a janitor during a period of contraction, but then subcontract with a janitorial service. Similarly, a ball bearing company may lay off a secretary and contract out to a clerical services company. Nothing has substantively changed in the process, the same jobs are performed, but the manufacturing sector in which the janitor or secretary used to work have registered a decrease in employment while the service sector has registered an increase. Because statistics on employment rely upon definitions based on industries and not occupations, this process results in a reduction of manufacturing employment and an increase in service employment. In terms of services employment, therefore, this process is something of a statistical mirage. Were definitions of services based on occupations and not industries, this factor would cease to be an explanation of services growth. Nonetheless, given the structure of national employment accounting systems, contracting out represents one reason why services employment has increased. This process serves as a reminder that the boundaries between services and manufacturing are often blurry (Walker, 1985).

Such issues point to the poorly understood issue of interregional trade in services. Evidence from the Italian Mezzogiorno (Martinelli, 1991) found that local firms purchased most of their externalized services from suppliers located in the north of the country. Similarly, Wood (1991) found that regions with inadequate pools of expertise were unlikely to enjoy the fruits of greater subcontracting. Thus, spatial discrepancies in the capacity to supply services are likely to be significant in shaping such patterns of trade.

Others have argued that the vertical disintegration and externalization thesis has been exaggerated. One line of thought holds that the demand for such services is satisfied through the growth of many small firms that satisfy new needs, often consisting of a single individual filling a highly specialized niche with customized needs, low overheads, and few economies of scale. As Table 6–1 reveals, the average size of establishments in U.S. producer services changed little between 1975 and 1989, but the number of them increased significantly. Such data indicate that services possess few internal economies of scale, a feature typically held to be true for flexible production systems. Thus, the degree to which firms engage in vertical disintegration, and the concomitant increases in subcontracting to service firms, is closely tied to related debates over trends in firm size and their effects on productivity and employment. The issue remains largely unresolved, with partisan advocates on both sides.

THE DIGITAL REVOLUTION AND THE CHANGING GEOGRAPHY OF PRODUCER SERVICES

As with all economic activities, the geographic distribution of services is highly uneven. The location of consumer services such as retail trade and personal services is relatively straightforward. Because most retail trade and personal service firms require close proximity to their market base, the distribution of such firms mirrors the distribution of population in general. For this reason, consumer services have been abundantly studied using central-place-theory approaches. The location of producer services, however, is altogether another story.

Any discussion of how flexible production technologies have altered the geography of services may begin by noting the similarity between the spatial distri-

Table 6–1: Establishment Population and Mean Size for U.S. Producer Services

	Number of Establishments		*Mean number of Employees*	
Sector	*1975*	*1989*	*1975*	*1989*
Transport services	14,607	40,887	9.29	8.33
F.I.R.E	372,245	529,258	11.45	12.85
Business services	123,024	263,592	15.90	18.02
Legal services	82,512	137,346	4.10	6.50
Managerial services	62,882	189,627	9.83	12.57
Total	655,271	1,160,704	11.16	13.07

Source: Compiled by William Beyers from *U.S. County Business Patterns*, 1975 and 1989.

bution of producer services and conglomerations of flexible production systems. Both tend to form highly agglomerated complexes over the landscape.

In the case of flexibly specialized firms, particularly electronics, this is most evident in the celebrated cases of Italy's Emilia-Romagna or California's Silicon Valley (Scott, 1983, 1988), which have been mythologized as dense, cooperative networks of small scale production, firms serving specialized niche markets using computerized production methods and just-in-time inventory systems.

The single most important consideration to take into account in studying the location of producer services is their tendency to agglomerate in large urban areas, including those in the nation's industrial heartland (e.g., Chicago, New York), which lowers their external transaction costs (O'hUallachain, 1989; O'hUallachain and Reid, 1991). Three powerful sets of reasons may be identified behind the incentive to cluster. First, metropolitan locations allow producer service firms to maximize their accessibility to clients and suppliers. Although transport costs are negligible, other benefits of agglomeration are critical. Firms can minimize their input and output costs by locating near their market base, which generally consists of other firms, including the headquarters of manufacturing firms. Advertising firms, for example, must have quick access to a variety of clients (hence the formation of specialized districts such as Madison Avenue in Manhattan), while securities firms may have need of rapid availability of specialized computer software firms or tax attorneys. Often such interconnections are mediated through face-to-face interaction (i.e., meetings of executives), which can be accomplished most profitably only in large urban areas. Face-to-face interaction allows firms to negotiate the details of particular transactions quickly and with security. Most managers spend most of their work time in meetings with suppliers or clients, and telecommunications are a poor substitute for face-to-face meetings unless the information is routine and standardized. In large cities, information often circulates in very narrow, specialized channels. By maximizing access to information, agglomeration reduces uncertainty.

Second, locations in large cities maximize producer services firms' access to the urban labor pool. This criterion is important because producer services tend to be labor intensive and because they often have need of specialized, skilled, white collar personnel who are not readily available in smaller cities.

Third, large urban areas offer sufficiently large markets to allow the existence of specialized service firms that would fail to find a client base large enough to sustain them in small cities. By spreading their costs over many clients, these firms thrive when otherwise they would not.

Fourth, like manufacturing firms, producer services have need of a well-developed urban infrastructure, largely to minimize transportation and communications costs, which tends to be found most commonly in large cities. The infrastructural requirements of service firms are quite different from their industrial counterparts. For example, because many producer service companies must send their executive or sales and marketing representatives to other cities, air transportation is critical. Similarly, many producer service firms require high-capacity telecommunications systems, including telephone lines, fiber optics cables, satellite earth stations, and microwave facilities (Warf, 1989). Some newly constructed "smart" office buildings include these facilities in their design (Moss and Dunau, 1986). Because of the overwhelming tendency of producer services to agglomerate in large cities, the rapid growth of business and financial services in the 1980s was a primary factor in reconstructing central city landscapes, including the rapid growth of downtown office complexes and widespread gentrification of inner city neighborhoods. In contrast to manufacturing, in which subcontracting linkages are increasingly inter-regional or international (Holmes, 1986), the backward linkages of services tend to be highly localized. In short, producer services have not followed manufacturing into the celebrated "new industrial spaces."

From an historical perspective, it is important to note that the location of producer services is heavily influenced by previous geographies created during earlier rounds of production. The clients and infrastructure of one region may reflect the types of production found there previously. For example, in both New York and London the complex of banking arose in large part because many merchant banking firms had close ties to the shipping industry. Further, the locational needs of producer services will vary from industry to industry. The geography of legal services, for example, is heavily affected by the presence of the federal government in Washington, D.C., where most large law firms have branch offices. London serves as even more of a legal services center for the U.K. than does Washington for the U.S. The location of securities firms in Manhattan is accentuated by the presence of the New York Federal Reserve there, which is a major player in the bond market. Finally, changes in international trade will shape the domestic geography of producer services. Miami has become a major banking center in large part because of its extensive ties to Latin American nations. Services in Vancouver, Seattle, and Los Angeles have grown quickly because of those cities' linkages to Pacific Rim trade and capital markets.

The tendency of producer services (particularly skilled headquarter and administrative functions) to agglomerate in a few selected metropolitan areas, the mobility of financial capital, the bifurcation between skilled and unskilled occupations, and contrasts between service and manufacturing wage levels have stimulated a lively debate about their capacity to initiate or enhance regional development (Coffey and Polèse, 1989; O'Farrell and Hitchens, 1990a, 1990b; Marshall and Wood, 1992). Clearly services are not an easy substitute for the loss of manufacturing jobs, a panacea for deindustrialization (Harrington and Lombard, 1989). Yet large numbers of the middle class throughout the industrialized world, as well as parts of the Third World, make a successful living in services. Thus, services reproduce the insistent tendency of capitalism to generate socially and spatially uneven development, neither eliminating poverty as early post-industrial theorists speculated nor breeding a new underclass tied to low-paying, dead-end jobs.

In the late twentieth century, the geography of producer services has undergone a remarkable transformation as the emergence of flexible production systems has led to widespread digitization of information and the employment of telecommunications systems. By the mid-1980s, with the conversion of office systems from analogue to digital form largely complete, many firms began to integrate their computer systems with telecommunications (Hepworth, 1990). As the cost of computing capacity dropped rapidly, communications became the largest bottleneck for information-intensive firms such as banks, securities brokers, and insurance companies. Numerous corporations, especially in financial services, invested in new communications technologies such as microwave and fiber optics. To meet the growing demand for high volume telecommunications, telephone companies upgraded their copper cable systems to include fiber optics lines, which allow large quantities of data to be transmitted rapidly, securely, and virtually error-free.

As data were converted from analogue to digital form, computer services merged with telecommunications. This phenomenon can be seen in no small part as an aftershock of the micro-electronics revolution and the concomitant switch from analogue to digital information formats. The digital format suffers less degradation over time and space, is much more compatible with the binary constraints of computers, and allows greater privacy (Akwule, 1992). In response to the growing demand for international digital data flows beginning in the 1970s, the United Nation's International Telecommunications Union introduced the Integrated Service Digital Network (ISDN) to harmonize technological constraints to data flow among its members (Akwule, 1992). ISDN has since become the standard model of tele-communications in Europe, North America, and elsewhere.

Such an event has produced an historic transformation of the ways in which information is collected, processed, and circulates, forming what Castells (1989) labels the "informational mode of production." In large part, such systems have contributed to the globalization of services (Dicken, 1992). The increasing reliance of business services, as well as numerous multinational manufacturing firms, upon telecommunications to relay massive volumes of information through international networks has made telecommunications a fundamental part of regional and national attempts to generate a comparative advantage in the information sector (Gillispie and Williams, 1988). The rapid deployment of such technologies reflects a conjunction of factors, including: the increasingly information intensive nature of commodity production in general (necessitating ever larger volumes of technical data and related inputs on financing, design and engineering, marketing and so forth); the spatial separation of production activities in different nations through globalized subcontracting networks; decreases in price and the elastic demand for communications; the birth of new electronic information services (e.g., on-line data bases, teletext, and electronic mail); and the high levels of uncertainty that accompany the international markets of the late twentieth century (to which the analysis of large volumes of data is a strategic response). The international expansion of these networks has raised issues of transborder data flows, intellectual property rights, copyright laws, etc., which have remained beyond the purview of traditional trade agreements. The computer networks that have made such networks technologically and commercially feasible offer users scale and scope economies, allowing spatially isolated establishments to share centralized information resources such as research, marketing and advertising, and management (Hepworth, 1986, 1990). Inevitably, such systems have profound spatial repercussions, reducing uncertainty for firms and lowering the marginal cost of existing plants, especially when they are separated from one another and their headquarters over long distances, as is increasingly the case.

Telecommunications allow not only new volumes of inter-regional trade in data services, but also in capital services. Banks and securities firms have been at the forefront of the construction of an extensive network of leased telephone networks (Langdale, 1989) and new technologies of satellites and fiber optics (Warf, 1989), giving rise to electronic funds transfer systems that have come to form the nerve center of the international financial economy, allowing banks to move capital around at a moment's notice, arbitraging interest rate differentials, taking advantage of favorable exchange rates, and avoiding political unrest (Langdale, 1985, 1989; Warf, 1989). Citicorp, for example, erected its Global Telecommunications Network to allow trade of $200 billion daily in foreign exchange markets around the world. Such networks give banks an ability to move money, by some estimates, more than $1.5 trillion daily (*Insight*, 1988), around the globe at stupendous rates. Subject to the process of digitization, information and capital become two sides of the same coin. In the securities markets, global telecommunications systems have also facilitated the emergence of the twenty-four-hour trading day, linking stock markets through the computerized trading of stocks. For example, in 1988, Reuters and the Chicago Mercantile Exchange announced the formation of Globex, an automated commodities trading system. In 1993, the New York Stock Exchange began the move to a twenty-four-hour-per-day trading system. Internationally, these changes created what Strange (1986) calls a "global gambling casino."

Concomitantly, there has occurred a profound change in the nature and function of finance, shifting the role of money-center banks away from raising capital for production (via stock markets) and hedging risks for commodities (via the futures markets) toward a variety of newer, more profitable functions such as foreign exchange (Pecchioli, 1983; Walter, 1988). The volatility of trading, particularly in stocks, has also increased as hair-trigger computer trading programs allow fortunes to be made (and lost) by staying microseconds ahead of (or behind) other markets. It is vital to note that heightened volatility, or the ability to switch vast quantities of funds over enormous distances, is fundamental to these capital markets. Speculation is no fun when there are no wild swings in prices.

The most readily evident geographic repercussions of this process have been the growth of "world cities," notably London, New York, and Tokyo (Moss, 1987a, 1987b; Sassen, 1991), each of which seems to be more closely attuned to the rhythms of the global economy than the nation-state in which it is located. London, for example, boomed under the impetus of the Euromarket in the 1980s, and has become detached from the rest of Britain (Thrift, 1987). Similarly, New York rebounded from the crisis of the mid-1970s with a massive influx of petrodollars and new investment funds (i.e., pension and mutual funds) that sustained a prolonged bull market on Wall Street. Tokyo, the epicenter of the gargantuan Japanese financial market, is likely the world's largest center of capital accumulation, with one-third of the world's stocks by volume and 12 of its largest banks by assets (Masai, 1989). In each, a large agglomeration of banks and related firms generates well-paying jobs. In each, soaring incomes for a wealthy stratum of traders and professionals has sent real estate prices soaring. It is important to note that urban specialization based on finance is not an historically new phenomenon (Amsterdam was the Wall Street of the seventeenth century), but the magnitude and rapidity of change in such markets is without precedent. Even with an eight-hour trading day, such centers are firmly connected by the global filiere of finance capital (Figure 6–2). With pressures building for a shift to the twenty-four-hour trading day, they are likely to be even more heavily connected.

However, telecommunications also threaten the agglomerative advantages of large urban agglomerations, particularly the advantages of face-to-face communications (as intimated long ago by the post-industrial thesis). For example, the National Associated Automated Dealers Quotation System (NASDAQ) has emerged as the world's fourth largest stock market. Unlike the New York, London, or Tokyo exchanges, NASDAQ lacks a trading floor, connecting 1/2 million traders worldwide through telephone and fiber optic lines. Similarly, Paris, Belgium, Spain, Vancouver, and Toronto all recently abolished their trading floors in favor of screen-based trading.

A second geographic manifestation of the new, hypermobile capital markets has been a surge of offshore banking, a reflection of the broader shift from traditional banking services (loans and deposits) to lucrative nontraditional functions, including debt repackaging, foreign exchange transactions, and cash management (Walter, 1988). The growth of offshore banking, inevitably in response to favorable tax laws, has stimulated banking in such previously unlikely places as Panama, Bahrain, and the Netherlands Antilles. Given the extreme mobility of finance capital and its increasing separation from the geography of employment, offshore banking can be expected to provide relatively few benefits for the nations in which it occurs. Roberts (1994), for example, illustrates the case of the Cayman Islands, now the world's fifth largest banking center in terms of gross assets, where 538 foreign banks employ only 1,000 people (less than two apiece). Such centers, typically in the margins of the world economy, represent attempts to take advantage of the world's uneven topography of

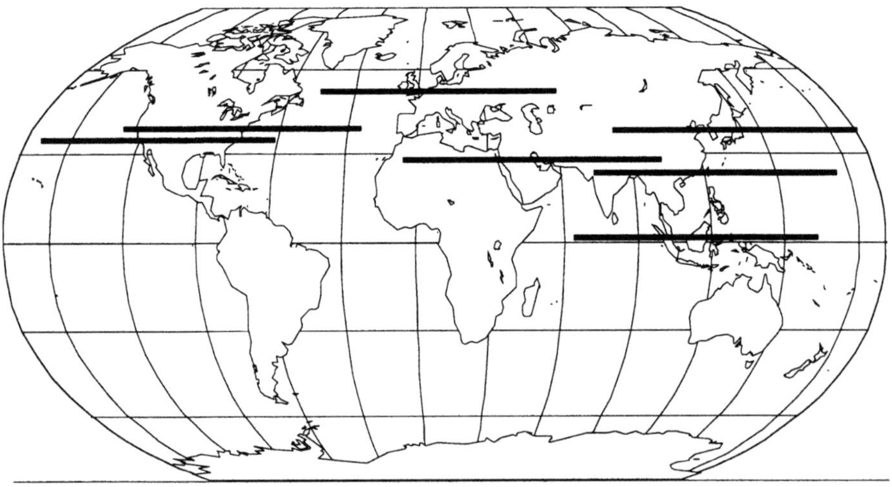

Figure 6–2: Trading Hours of Major World Financial Centers.

regulation. As the technological barriers to capital flow have declined, the importance of political ones has risen concomitantly.

A third manifestation of these systems concerns the globalization of office functions, particularly back offices. Back offices perform many routinized clerical functions such as data entry of office records, telephone books, library catalogues, stock transfers, processing of payroll or billing information, bank checks, insurance claims, magazine subscriptions, and airline frequent flyer coupons. These tasks involve use of unskilled or semi-skilled labor, primarily women, and frequently operate on a twenty-four-hour-per-day basis. Back offices have few of the interfirm linkages associated with headquarters activities and require extensive data processing facilities, reliable sources of electricity, and sophisticated telecommunications networks.

Historically, back offices have located adjacent to headquarters activities in downtown areas to insure close management supervision and rapid turnaround of information (Nelson, 1986). However, under the impetus of rising central city rents and shortages of sufficiently qualified (i.e., computer literate) labor, many service firms began to uncouple their headquarters and back office functions, moving the latter out of the downtown to cheaper locations on the urban periphery. Most back office relocations, therefore, have been to suburbs (Moss and Dunau, 1986). Recently, given the increasing locational flexibility afforded by satellites and a growing web of inter-urban fiber optics systems, back offices have begun to relocate on a much broader, continental scale. Under the impetus of new telecommunications systems, many clerical tasks have become increasingly footloose and susceptible to spatial variations in production costs. For example, several firms fled New York City in the 1980s:

- American Express moved its back offices to Salt Lake City, Utah (850 jobs) and Phoenix, Arizona;

- Citicorp shifted its Mastercard and Visa divisions to Tampa, Florida (600 jobs), and Sioux Falls, South Dakota, and moved its data processing functions to Las Vegas, Nevada; Buffalo, New York; Hagerstown, Maryland; and Santa Monica, California;

- Citibank moved its cash management services to New Castle, Delaware;

- Chase Manhattan housed its credit card operations in Wilmington, Delaware;

- Hertz relocated its data entry division to Oklahoma City, Oklahoma;

- Avis went to Tulsa, Oklahoma;

- Dean Witter moved its data processing facilities to Dallas, Texas;

- Metropolitan Life repositioned its back offices to Greenville, South Carolina; Scranton, Pennsylvania; and Wichita, Kansas;

- Deloitte Haskins Sells relocated its back offices to Nashville, Tennessee;

- Eastern Airlines shifted to Miami, Florida (Warf, 1993).

Internationally, this trend has taken the form of the offshore office (Wilson, 1991). The primary motivation for offshore relocation is low labor costs, although other considerations include worker productivity, skills, turnover, and benefits. Offshore offices are established not to serve foreign markets, but to generate cost savings for U.S. firms by tapping Third World labor pools where wages are as low as one-fifth of those in the U.S. Notably, many firms with offshore back offices are in industries facing strong competitive pressures to enhance productivity, including insurance, publishing, and airlines. Offshore back office operations remained insignificant until the 1980s, when advances in telecommunications such as transoceanic fiber optics lines made possible greater locational flexibility just when the demand for clerical and information processing services grew rapidly. However, the offshoring of back offices probably only represents a small fraction of the 250,000 U.S. workers employed in the data processing industry. The mechanics of offshore offices are relatively simple. Such functions may be either subsidiaries of multinational firms or operate under contract with U.S.-based businesses. Inputs, usually documents or magnetic tapes, are sent by air to offshore processing facilities. After processing, generally a few days at most, the results are returned via air couriers, dedicated telephone line, satellite, or fiber optics. Thus, the capital investments in such operations are minimal.

Offshore back offices have sprouted up rapidly in various parts of the world. Several New York-based life insurance companies have erected back office facilities in Ireland, with the active encouragement of the Irish government (Lohr, 1988). Often situated near Shannon Airport (at which all international flights to Ireland must first land), they move documents in by Federal Express and the final product back via satellite or the TAT-8 fiber optics line that connected New York and London in 1989 (Figure 6–3). Despite the fact that back offices have been there only a few years, Irish development officials already fret, with good reason, about potential competition from Greece and Portugal. Likewise, the Caribbean has become a particularly important locus for American back offices, partly due to the Caribbean Basin Initiative instituted by the Reagan Administration and the guaranteed access to the U.S. market that it provides. Most back offices in the Caribbean have chosen Anglophonic nations, particularly Jamaica and Barbados. American Airlines paved the way in the Caribbean through its subsidiary Caribbean Data Services (CDS), which began when a data processing center moved from Tulsa to Barbados in 1981. In 1987, CDS opened a second office near Santo Domingo, Dominican Republic, where wages are one-half as high as Barbados. Thus, the same flexibility that allowed back offices to move out of the U.S. can be used against the nations to which they relocate. Asia, with massive reserves of low cost labor, would also seem a likely place in which back offices may relocate. For example, Manila, the Philippines, is home to over 40 firms (Wilson, 1991).

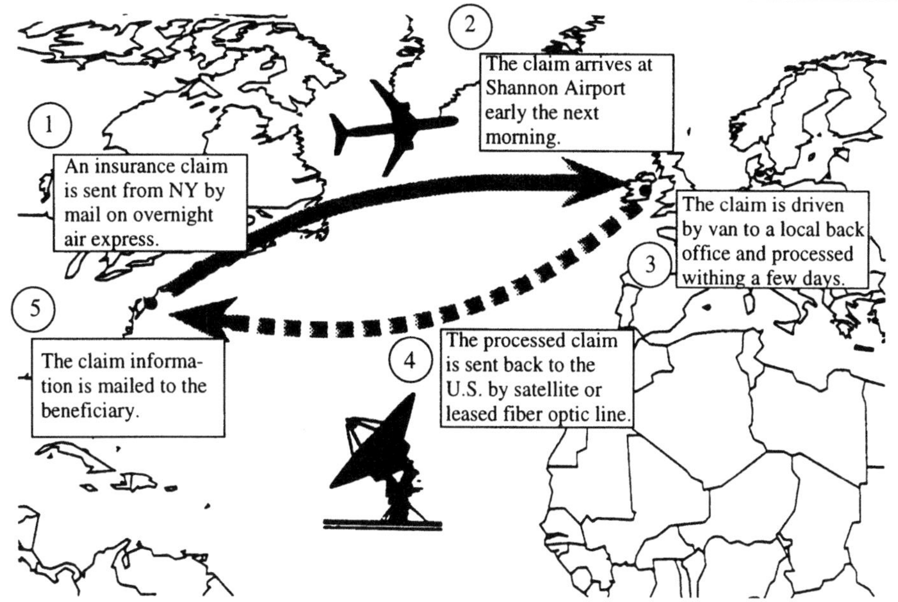

Figure 6–3: Mechanics of Back Office Relocation to Ireland.

CONCLUDING THOUGHTS

From the remarks offered above, it is readily evident that services and flexible production systems interact in highly complex ways. On the one hand, as Wood (1991) and others have noted, services play an important role in the growing flexibility of capitalist enterprises to deal with much higher levels of uncertainty and rapidity of change than in the past. Given that many producer service inputs are sporadic in nature, many firms have externalized such functions, although the growth of small, independent establishments cannot be denied. On the other hand, services have incorporated numerous flexible production technologies that have radically realigned the nature of work within producer service industries as well as their geographical location. Foremost among these changes has been the digitization of information and the introduction of telecommunications, which simultaneously reinforce and challenge the enormous pull of metropolitan agglomeration economies. With the digital revolution, services can no longer be so easily dismissed as parasitic or locally consumed activities, for mounting evidence indicates that they are traded interregionally and internationally. The widespread adoption of flexible production technologies has contributed both to the centralization of skilled, high wage, white collar jobs in "global cities" and to the decentralization of relatively unskilled, standardized functions.

As part of the broad sea-change from Fordist production regimes to the globalized world of flexible accumulation, it is clear that finance capital acquires a dynamic qualitatively different from that of manufacturing. The increased fluidity of finance capital on a global basis, particularly following the worldwide wave of deregulation unleashed in the 1970s, revolves around the introduction of electronic funds transfer systems, which give banks, securities, and insurance firms markedly greater freedom over their locational choices. Such mobility is clearly evident, for example, in recent studies of back offices and offshore banking. In dramatically reducing the circulation time of capital, telecommunications have linked far flung places together through networks in which billions of dollars move instantaneously across the globe, creating a geography without transport costs, as attested by the current status of cities such as London, New York, Tokyo, Toronto, Singapore, and Frankfurt, as well as Bahrain, Panama, and the Cayman Islands.

It is vital to note, therefore, that contrary to early, simplistic expectations that telecommunications would "eliminate space," rendering geography meaningless through the effortless conquest of distance, such systems in fact produce new rounds of unevenness, forming new geographies that are imposed upon the relics of the past. Telecommunications both reflect and transform the topologies of capitalism, creating nested hierarchies of spaces technically articulated in the architecture of computer networks (Hepworth, 1986, 1990). Far from eliminating differences between places, telecommunications permit the exploitation of those differences.

Finally, while the interconnections between services and flexible production systems have been stressed here, their autonomy is also worth noting. The service economy emerged in the late 19th century, long, long before the birth of flexibility. As discussed earlier, the reasons for the rapid growth of services lay in the changing incomes, demographic structure, and degrees of public intervention in western societies. Equally evident, however, is that services would not be exchanged to the degree that they are today without benefit of flexible production systems. In short, these two arenas of economic activity, each of which embraces a variety of production techniques, are contingent but mutually reinforcing systems, reflecting and reinforcing the evolution of commodity production.

REFERENCES

Akwule, R. (1992). *Global Telecommunications: The Technology, Administration, and Policies*. Boston: Focal Press.

Amin, A. and Robins, K. (1990). "The Re-emergence of Regional Economies? The Mythical Geography of Flexible Accumulation," *Environment and Planning D: Society and Space* 8, pp. 7–34.

Baran, B. (1985). "Office Automation and Women's Work: The Technological Transformation of the Insurance Industry." In M. Castells (Ed.), *High Technology, Space, and Society*. Beverly Hills: Sage.

Beyers, W. and Alvine, M. (1985). "Export Services in Post-Industrial Society," *Papers of the Regional Science Association* 57, pp. 33–46.

Brunn, B. and Leinbach, T. (Eds.) (1991). *Collapsing Space and Time*. London: Harper Collins.

Castells, M. (1989). *The Informational City*. Oxford: Basil Blackwell.

Coffey, W. and Bailly, A. (1991). "Producer Services and Flexible Production: An Exploratory Analysis," *Growth and Change* 22, pp. 95–117.

Coffey, W. and Polèse, M. (1989). "Producer Services and Regional Development: A Policy-Oriented Perspective," *Papers of the Regional Science Association*, 67, pp. 13–27.

Daniels, P. (1991). "Some Perspectives on the Geography of Services," *Progress in Human Geography* 15, pp. 37–46.

Dicken, P. (1992). *Global Shift: The Internationalization of Economic Activity*. (2nd ed.) New York: Guilford Press.

Gillispie, A. and Williams, H. (1988). "Telecommunications and the Reconstruction of Comparative Advantage," *Environment and Planning A* 20, pp. 1311–21.

Ginzberg, E. and Vojta, G. (1981). "The Service Sector of the U.S. Economy," *Scientific American* 244, pp. 48–65.

Goe, W. (1990). "Producer Services, Trade and the Social Division of Labour," *Regional Studies* 24, pp. 327–42.

———. (1991). "The Growth of Producer Services Industries: Sorting through the Externalization Debate," *Growth and Change* 22, pp. 118–41.

Graham, J. et al. (1988). "Restructuring in U.S. Manufacturing: The Decline of Monopoly Capitalism," *Annals of the Association of American Geographers* 78, pp. 473–90.

Hall, P. and Preston, P. (1988). *The Carrier Wave: New Information Technology and the Geography of Innovation. 1846–2003*. London: Unwin Hyman.

Harrington, J. and Lombard, J. (1989). "Producer-Service Firms in a Declining Manufacturing Region," *Environment and Planning A* 21, pp. 65–79.

Harrison, B. and Bluestone, B. (1988). *The Great U-Turn: Corporate Restructuring and the Polarizing of America*. New York: Basic Books.

Harvey, D. (1989). *The Condition of Postmodernity*. Oxford: Basil Blackwell.

Hepworth, M. (1986). "The Geography of Technological Change in the Information Economy," *Regional Studies* 20, pp. 407–24.

_____ . (1990). *Geography of the Information Economy*. London: Guilford Press.

Holly, B. (1987). "Regulation, Competition, and Technology: The Restructuring of the U.S. Commercial Banking System," *Environment and Planning A* 19, pp. 633–52.

Holmes, J. (1986). "The Organization and Locational Structure of Production Subcontracting." In A. Scott and M. Storper (Eds.), *Production, Work, Territory: The Geographical Anatomy of Industrial Capitalism*. Boston: Allen and Unwin, pp. 80–106.

Insight. (1988). "Juggling Trillions on a Wire: Is Electronic Money Safe?" Feb. 15, pp. 38–40.

Kakabadse, M. (1987). *International Trade in Services: Prospects for Liberalisation in the 1990s*. London: Croom Helm.

Langdale, J. (1985). "Electronic Funds Transfer and the Internationalisation of the Banking and Finance Industry," *Geoforum* 16, pp. 1-13.

_____. (1989). "The Geography of International Business Telecommunications: The Role of Leased Networks," *Annals of the Association of American Geographers* 79, pp. 501–22.

Lohr, S. (1988). "The Growth of the Global Office," *New York Times*, Oct. 18, p. D1.

Marshall, J. (1982). "Linkages Between Manufacturing Industry and Business Services," *Environment and Planning A* 14, pp. 1523–40.

_____. (1988). *Services and Uneven Devleopment*. Oxford: Oxford University Press.

Marshall, J. and Wood, P. (1992). "The Role of Services in Urban and Regional Development: Recent Debates and New Directions," *Environment and Planning A* 24, pp. 1255–70.

Martinelli, F. (1991). "Branch Plants and Services Underdevelopment in Peripheral Regions: the Case of Southern Italy." In P. Daniels and F. Moulaert (Eds.), *The Changing Geography of Advanced Producer Services*. London: Bellhaven, pp. 15–29.

Masai, Y. (1989). "Greater Tokyo as a Global City." In R. Knight and G. Gappert (Eds.), *Cities in a Global Society*. Newbury Park, CA: Sage, pp. 153–63.

Moss, M. (1987a). "Telecommunications and International Financial Centers." In J. Brotchie, P. Hall, and P. Newton (Eds.), *The Spatial Impact of Technological Change*. London: Croom Helm, pp. 116–29.

______. (1987b). "Telecommunications, World Cities, and Urban Policy," *Urban Studies* 24, pp. 534–46.

Moss, M. and Dunau, A. (1986). "Offices, Information Technology, and Locational Trends." In J. Black, K. Roark, and L. Schwartz (Eds.), *The Changing Office Workplace*.Washington, DC: Urban Land Institute, pp. 171–82.

Nelson, K. (1986). "Labor Demand, Labor Supply and the Suburbanization of Low-wage Office Work." In A. Scott and M. Storper (Eds.), *Production, Work, Territory*. Boston: Allen Unwin, pp. 149–71.

Noyelle, T. And Dutka, A. (1988). *International Trade in Business Services*. Cambridge, MA: Ballinger.

O'Farrell, P. and D. Hitchens. (1990a). "Producer Services and Regional Development: Key Conceptual Issues of Taxonomy and Quality Measurement," *Regional Studies* 24, pp.163–72.

_____. (1990b). "Producer Services and Regional Development: A Review of Some Major Conceptual and Policy and Research Issues," *Environment and Planning A* 22, pp. 1141–54.

O'hUallachain, B. (1989). "Agglomeration of Services in American Metropolitan Areas," *Growth and Change* 20, pp. 34–49.

O'hUallachain, B. and Reid, N. (1991). "The Location and Growth of Business and Professional Services in American Metropolitan Areas, 1976–1986," *Annals of the Association of American Geographers* 81, pp. 254–70.

Pecchioli, R. (1983). *The Internationalization of Banking: The Policy Issues*. Paris: OECD.

Quinn, J.; Baruch, J.; and Paquette, P. (1987). "Technology in Services," *Scientific American* 257, (6), pp. 50–58.

Roberts, S. (1994). "Fictitious Capital, Fictitious Spaces: the Geography of Offshore Financial Flows." In S. Corbridge, R. Martin, and N. Thrift (Eds.), *Money, Power, and Space.* Oxford: Basil Blackwell, pp. 91–115.

Sassen, S. (1991). *The Global City: New York, London, Tokyo.* Princeton, NJ: Princeton University Press.

Sayer, A. (1984). (2nd ed., 1992). *Method in Social Science: A Realist Approach.* London: Routledge.

Scott, A. (1983). "Location and Linkage Systems: A Survey and Reassessment," *Annals of Regional Science* 17, pp. 1–39.

_____. (1988). "Flexible Production Systems and Regional Development: The Rise of New Industrial Spaces in North America and Western Europe," *International Journal of Urban and Regional Research* 42, pp. 171–86.

Scott, A. and Storper, M. (1987). "High Technology Industry and Regional Development: A Theoretical Critique and Reconstruction," *International Social Science Journal* 112, pp. 215–32.

Shrivastava, P. (1983). "Strategies for Coping with Telecommunications Technology in the Financial Services Industry," *Columbia Journal of World Business* 18, pp. 19–25.

Stabler, J. and Howe, E. (1988). "Service Exports and Regional Growth in the Postindustrial Era," *Journal of Regional Science* 28, pp. 303–15.

Stanback, T. (1979). *Understanding the Service Economy.* Baltimore: Johns Hopkins University Press.

Storper, M. and Christopherson, S. (1987). "Flexible Specialization and Regional Industrial Agglomeration: The Case of the U.S. Motion Picture Industry," *Annals of the Association of American Geographers* 77, pp. 104–17.

Storper, M. and Walker, R. (1989). *The Capitalist Imperative: Territory, Technology, and Industrial Growth.* Oxford: Basil Blackwell.

Strange, S. (1986). *Casino Capitalism.* Oxford: Basil Blackwell.

Thrift, N. (1987). "The Fixers: The Urban Geography of International Commercial Capital." In J. Henderson and M. Castells (Eds.), *Global Restructuring and Territorial Development.* Beverly Hills: Sage Publications, pp. 203–33.

Tuttle, D. (1987). "Whether Your Business is Tourism or Not, Tourism is Your Business," *Business America*, February 16, pp. 3–10.

Walker, R. (1985). "Is There a Service Economy? The Changing Capitalist Division of Labor," *Science and Society* (spring), pp. 42–83.

Walter, I. (1988). *Global Competition in Financial Services: Market Structure, Protection, and Trade Liberalization.* Cambridge, MA: Ballinger.

Warf, B. (1989). "Telecommunications and the Globalization of Financial Services," *Professional Geographer* 41, pp. 257–71.

_____ . (1993). "Back Office Dispersal: Implications for Urban Development," *Economic Development Commentary* 16, pp. 11–16.

Wilson, M. (1991). "Offshore Relocation of Producer Services: The Irish Back Office." Paper presented at the annual meeting of the Association of American Geographers, Miami.

Wood, P. (1991). "Flexible Accumulation and the Rise of Business Services," *Transactions of the Institute of British Geographers NS* 16, pp. 160–72.

7

INTER-FIRM COLLABORATION AND INDUSTRIAL COMPETITIVENESS

Brian Bosworth
Regional Technology Strategies, Inc., 1955 Massachusetts Avenue, Cambridge, MA 02140 USA

INTRODUCTION

At the high end of manufacturing markets, the end that will create wealth and ultimately jobs, things don't get made by companies anymore. Things get made, designed, engineered, manufactured, and distributed by groups of firms working together in new, dynamic, and multiple alliances and networks. For firms to be competitive in these markets, there is just *too* much to learn, *too* much to be able to do. They can't do it alone any longer. They have to learn to learn cooperatively.

Unlike many of the essays in this volume, my essay is based not on academic research, but on practical experience as a consultant. In the remainder of this chapter, I first stress the recent changes in the global competitive environment. This is followed by discussions of the importance of adding value and the seven dimensions of modernization. I conclude by speculating on appropriate public policies to assist modernization.

BASIC CHANGES IN THE WORLD ECONOMY

We begin by reviewing those basic changes in the world economy that have radically altered the competitive environment for U.S. manufacturers, changes that now drive best practice manufacturers of all sizes toward various forms of collaboration. It is helpful to understand this period of fundamental economic adjustment as involving change in markets, products, technology, and business structure. Some caution is necessary to avoid defining these changes as events. They are interrelated and dynamic trends, still unfolding and still responding to one another. Viewed collectively, they help to develop a deeper appreciation of the issue of competitiveness.

Changes in Markets: Segmentation, Globalization, and Volatility

Mass national markets have been replaced by segmented international markets. The revolution in information processing, transportation, and telecommunication technologies, propelled by divergent consumer tastes, has produced an increasingly niche-oriented marketplace. Rising affluence worldwide has contributed to highly articulated demand for consumer goods and, in turn, for the producer goods to make them. What used to be mass markets of undifferentiated demand have shattered into narrow fragments, each representing specialized demand for specialized products. Standardized products find little acceptance in this segmented demand environment.

This segmentation of market has been accompanied by an internationalization of the market. Notwithstanding periodic policy outbursts about "fair trade" and protectionism, the past 30 years of steady negotiation on tariffs and trade issues have resulted in the fundamental relaxation of political barriers to international trade. The aggressive, almost frenetic industrialization policies of other nations, especially in Asian nations that have long relied on exports to drive their economies, have paid off (at least for them). Businesses and governments in such Pacific Rim nations as Japan, Singapore, and Korea and in European nations like Germany, Denmark, and Italy have proactively sought to develop an export-based economy. Most manufacturing firms there of all sizes have learned what it takes to compete in international markets.

Intense foreign competition now pressures businesses that have always seen themselves as on top of the international market or isolated from it. Manufacturers in America can elect not to export their products, but they can't opt out of international competition. Goods produced in other countries now compete in virtually every corner of the American market. Many American firms are finding that if they have not honed their competitive edge in the export market, they cannot hold their own in the domestic market. In America, even the large firms have been slow to develop the ability to meet international standards of quality, delivery, reliability, and price. Small firms have lagged much further behind in learning to compete against global standards.

Another characteristic of the changing market environment for manufactured goods is the volatility of these markets. Because they are so narrow and so fragmented, they change very quickly; old ones go away and new ones emerge.

Firms seeking to be competitive in this market environment must be prepared to first find and then satisfy very narrow market segments. And they better do it quickly before the demand mutates to a new form and the market goes away. The markets are not necessarily close by and those that may be are attracting competitors from all over the world. A successful firm must be prepared to shift quickly to follow these changing markets and/or be very good at discovering new ones. It is hard for a small firm, without worldwide market sensing capabilities, to compete by anticipating market shifts and frequently smaller firms (and big ones as well) are led to new relationships with old customers and new alliances with firms with which they also compete.

Changes in Products: From Standard-Made to Tailor-Made

The flip side of these changes in markets is changes in products. Standard products have given way rapidly to customized products. Purchasers of standard, mass-produced goods that American manufacturers were very good at making were usually willing to accept products that came pretty close to meeting some of their needs. But niche demand requires highly differentiated products that precisely satisfy the very narrow application requirements of the buyer.

Customized products, therefore, must embody higher levels of skill, knowledge, and technology than did standardized products. Customized products usually require application of sophisticated manufacturing equipment organized into sophisticated systems and operated by sophisticated personnel.

As observed, niche markets are volatile markets and therefore customized products tend to have very short life cycles. They are often very new products, tailor-made for the buyer, or mature products, greatly specialized to particular applications. Such products do not compete principally on the basis of cost. They compete rather on the basis of quality, precision, and reliability of delivery as well as performance. Time is frequently more important than price to the buyer. Cost is determined by the maker but value can be established only by the buyer.

Customized products are much more highly valued and therefore a firm that gets better and better at customizing adds more and more value to the production process. The greater the degree to which a product can be made or delivered in a way that is customized to the needs of an individual customer, the greater the value to that customer and the more that customer will pay for it. Cost continues to be an important, while not determinant, factor in competitiveness. However, cost control cannot normally be found in the scale of production. Instead it must be found in design, in "manufacturability," in logistics of supply, delivery, and service, and in the quality of management.

It is not quite correct to see mass production as "disappearing" in the same sense that mass markets are. Some forms of mass production remain, altered by new approaches to materials planning and inventory management and new production relationships into what have been termed "lean production" models. Not all specialized products designed for niche markets are characterized by short production runs or very low volume. Sometimes the value added is precisely the ability to make and ship a whole lot of a particular item very quickly.

Changes in Industrial Technology: From Automated Single Purpose to Reprogammable Flexibility

These changes in markets are driven by and undergirded by fundamental changes in manufacturing technology. The advantage in manufacturing has changed over the past several years from being good at the routine to being good at the flexible and being able to change quickly.

Hard automation technology of the past several decades involved capturing "know-how" and building it into single-purpose machines and single-purpose systems. Flexibility was sacrificed for efficiency. However, newer manufacturing technology has aimed for flexibility and to accommodate a wider variety of customized products for the short-life niche markets. The technology of micro-electronic controls and re-programmable automation is aimed at economies of scope, not scale. Multiple products, each tailored for different applications, can be produced almost as cheaply (or sometimes even more cheaply) in combination off the same machine or system of machines than separately off different machines or different systems of machines. Computer numerically controlled machines and computer integrated manufacturing systems permit variety and flexibility of product without reducing efficiency.

This newer technology also reduces production time, improves quality, and slashes inventory. Computerized planning, inventory, and group technology systems can track multiple products through the factory while computer-based process control systems can assure uniform high levels of quality. Thus industrial process technology both underlies and propels the changes to narrower market segments and customized products.

It is not always obvious to manufacturers, especially smaller ones, that investing in these technologies is wise. Advanced manufacturing technologies are not some monolithic thing. They are discrete and separable. Some make sense for some companies in some situations. Some acquisitions are expensive, some are daunting from a management perspective and some assume skills that many workers do not have. Moreover, these technologies cannot be optimized in the absence of very difficult, parallel changes in several other dimensions. Still, the adoption of these technologies by competitors who can afford them, manage them, run them, and optimize them has dramatically altered the competitive environment for those who have not.

Changes in the Structure of Industrial Enterprise: The Rise of Agile Enterprise

Rapid changes in markets, products and processes are driving enormous changes in the structure of manufacturing establishments and the relationships among them. They are shrinking in size, shedding overhead, and stripping away embedded layers of management hierarchy.

Businesses must react quickly to rapid demand shifts in volatile niche markets. They must respond promptly to innovation in technology and to the demand for more specialized, higher quality and shorter-lived products. The need for speed and accuracy in that response is driving a general decentralization of organization.

Larger manufacturing companies are devolving into smaller, more autonomous business units. Headquarters' staffs are shrinking rapidly. Many firms that have had as many as 12 or even 15 layers of management between the chief executive and front-line supervisors are cutting back to six, or five, or even less. With good use of good information systems, managers who used to think they could properly supervise only a dozen employees are finding that they can effectively communicate with as many as 200.

A flatter structure within the manufacturing organization permits greater communication and cooperation among different divisions of the corporation. This in turn facilitates the increasingly sensitive, internal, and vertical communication needs which are crucial to the processing of the huge volume of information necessary to understand complex, volatile markets and technologies.

In many industries, these trends are contributing to a disintegration of production systems. Many end product firms are "hollowing-out," out-sourcing an increasing percentage of the component requirements. This encourages specialization among their suppliers. As special opportunities or special problems emerge, they are addressed not by rebuilding large production groups, but by ad hoc task forces, special consultants, outside service vendors, new, specialized suppliers and new alliances among customers, suppliers, collaborators and competitors.

THE IMPORTANCE OF ADDING VALUE

As these trends play out, the value of manufactured goods increasingly reflects the extent to which they embody advanced technologies, skills, and business practices which satisfy the specialized needs of segmented markets. Products that meet these exacting requirements have high value-added in design, engineering, manufacturing, and delivery. Value-added is the difference between what the firm pays for the raw material, parts, and components that it buys and the amount for which it sells its finished goods. The more the value-added on a per-employee basis, the more wealth is created by the enterprise and the greater the economic return to workers, managers and investors. Value can be added through the application of sophisticated process technologies and the smart work of highly skilled workers and managers. Value can be added in the design and engineering process by incorporating product features and capabilities for which the customer is willing to pay a premium. Value can be added in the delivery logistics by getting the product to the customer in the time, the sequence, and the mix that the customer needs. In other words, value-added is not strictly a matter of productivity; it also reflects quality and service. Value is not at all the same thing as cost; a firm cannot necessarily add more value simply by reducing cost. Cost is established by the producer. Value is determined strictly by the customer.

Those firms using relatively basic technologies, low-skilled workers and managers, and very traditional business practices are usually not going to be able to add a great deal of value to the raw materials and component goods with which they start. They will be able to compete only at the relatively low ends of their markets where products must meet far less exacting requirements, where they are more homogeneous, less differentiated and therefore tend to compete chiefly on the basis of price. On the other hand, those firms which figure out quicker than others how to adopt or adapt the highest levels of technologies, how to deploy very highly skilled workers into organizational systems that impel learning, and how to use business practices that promote customer response and quality products will be able to compete more aggressively at the highest ends of their markets.

From a public policy point of view, we want a manufacturing economy that is very good at producing the highest value goods and services. Only by producing high-value goods can our manufacturers pay the high wages and salaries that will contribute to a rising standard of living. If our firms are slow in gaining the ability to produce high value goods and therefore are forced to compete at lower levels of value-added, they will be pressured by competitive forces to pay lower rather than higher wages because, at relatively low levels of value-added, they are compelled to compete more on the dimension of price not on the dimension of value. Often, that means they will seek out and migrate to low cost production environments in other parts of the world or they will exert downward pressure on production costs here, downward pressure which can only have the consequence of reducing our standard of living. If the public policy objective of economic development strategies is a rising standard of living, that can be achieved only if our manufacturers try to compete at higher levels of value-added.

THE DIMENSIONS OF MODERNIZATION

If the overall goal of regional economic development from the perspective of the firm is competitiveness and if the kind of competitiveness we seek is the ability to compete successfully at high levels of value-added, then we must understand that this is a multidimensional process. There is no single solution to the problems of modernization. If assistance programs are to be effective, managers of those programs must get good at understanding and acting upon the connections and interdependencies of change in the modernizing firm. There are seven key dimensions of change for modernizing firms.

Technology

It is usually not possible to produce highly customized, quality-oriented products for narrow and volatile market segments without employing sophisticated manufacturing technologies. Most frequently find advanced technology will optimize such functions as:

- design and engineering where computer-aided machines and systems can perform complex procedures quickly and accurately and can provide digital representation of input requirements and link the output directly to manufacturing operations;

- materials and requirements planning where new systems can optimize cost and utilization of raw materials, labor and time;

- fabrication, machining, and assembly where flexible cells or systems, numerically or computer numerically controlled materials preparation devices, and working, cutting, and assembly devices can dramatically improve the speed, accuracy and multi-use capabilities of machines and operations;

- materials handling and inventory management systems which can reduce cost and rationalize scheduling and inventory flows;

- inspection and testing using automated, sensor-based inspection and testing equipment for incoming, in-process or assembled goods; and,

- communications and control where micro-electronic and communications systems link manufactures internally and with other firms elsewhere.

Within each area of this rough categorization, there are scores of choices that a firm must make when deciding to upgrade machines or systems. Strictly within the framework of the technological choice (ignoring questions related to skills required, cost, etc.) there are problems of compatibility with existing machines and systems and with the technology of supplies and customers. Many, perhaps most, small manufacturers do not have an engineer on staff to help them through the selections and the owners and managers often mistrust the equipment and software vendors.

Why don't smaller firms deploy the kind of advanced technology that will help them compete in high end markets? In other words, why have firms made consistent decisions, year after year, to not purchase and install manufacturing technologies with which they are at least generally familiar and which purport to have resulted in considerable efficiencies and created capacities for higher valued-added production? The limited evidence that is available here suggests that the prime obstacle is money. In 1991, the Bureau of the Census of the U.S. Department of Commerce published the results of its national survey of about 10,000 manufacturers in SIC codes 34–38. When the respondents were asked to specify their most significant problems in acquiring or using technologically advanced equipment and software over the preceding three years, the leading problem identified, in every category of technology, was the overall cost. The second most frequent problem cited was the cost of education and training and the third was the lack of financial justification. When the firms were asked to look ahead over the next three years and identify the anticipated barriers, they again cited cost far more frequently than anything else.

Interestingly, when these firms were asked to rank the most important benefits they derived from the new technology they had installed, the gain cited most frequently (except in reference to materials handling equipment) was quality improvement. Cost savings was generally the second most frequently cited benefit with flexibility increases, and lead time reductions also being cited frequently.

The survey did not ask firms who had not installed new technology why they had not. Anecdotally, when firms are asked that question they tend to respond that they do not need it or their customers don't ask for it. This again suggests that small firms don't install new technology because they see it as expensive and because their "business strategy" doesn't require it.

Advanced Business Management Practices

Smaller firms are not likely to cope successfully with the complexity of interrelated and, often, simultaneous change in all the dimensions outlined above without utilizing advanced business management practices. Adapting long-range strategic planning, figuring out how to slash inventories and keep them low, planning for material requirements, implementing total quality management systems, and other such practices to small firms with scarce resources is daunting but necessary. These business practices are essential to gain value from the advanced technology.

Markets

As discussed above, high-value markets are international in scope, highly segmented in scale, and extremely volatile in demand. Small manufacturers who have been accustomed to more local, homogenous, and stable markets often find they must gain new skills in market research, analysis and planning.

Smaller firms have generally defined themselves by what they make; that has been their primary mission. As the markets evolve, most are working harder to find and satisfy customers who want to buy those things and that occupies more time, but it typically is still a secondary function. Very few see themselves as having a *core skill* in understanding markets. A lot of small firms do not actively search for new markets at all. They depend on orders from an established customer base. If a lot of orders come in, business is "good." If orders fall off, business is "bad."

The relationship between market research, analysis, and planning capability is direct and immediate. If customers do not want products which require advanced technology, firms are not going to install advanced technology. If they do adopt advanced technologies, they must learn to optimize the value of it by developing new markets where its application will add value.

Work Organization

High-value production demands flexible enterprises that can respond quickly to customer requirements for quality, reliability, precision, and speed of delivery. Centralized and hierarchical work organizations retard achievement of these objectives. Many modernizing firms are shifting quickly toward flexible or "high performance" work organizations. Typical attributes of these re-organized work systems are severely flattened management structures, employee work teams with significant autonomy, job rotation, and employee involvement in problemsolving.

The key to competitiveness is organization. New forms of external organization are crucial as we will see in a minute. But the first place to begin is with internal organization. The object is to eliminate internal barriers to learning. In high value manufacturing, the firm just can't afford these barriers. Reorganizing work means getting rid of them.

Skills Enhancement

Managers and workers in the modernizing firm often need new occupational, technical, and problemsolving skills to effectively deploy advanced technology and flexible work organizations. Many firms find that developing the advanced technical skills frequently requires upgrading basic literacy and the numeric and communication skills of their employees.

Finance

Exploring new markets, acquiring new technologies, and investing in training requires money which smaller firms typically cannot raise in public markets or internally through retained earnings. Therefore, the modernization process leads frequently to greater demands for credit.

In the late 1980s, federal regulatory agencies imposed stringent new underwriting criteria on banks and other commercial and industrial lending institutions. These have led to more restrictive requirements on collateral, debt-asset ratios, and other lending terms that many smaller firms in particular have problems meeting.

Some development practitioners have suggested that the real issue here is financial expertise within the small firm. Few small manufacturing companies have acquired sophisticated capacity to manage their credit and capital needs. They have not previously been in a competitive position that stresses their capital management capacity and they have been slow to respond to the changing financial environment.

Inter-Firm Cooperation

Virtually all firms discover that modernization requires new skills and new mechanisms for managing their relationships with other companies—be they customers, suppliers, competitors, or other firms tightly linked in marketing and production clusters. Smaller firms are often inexperienced at consortial action in pursuing joint solutions to common problems and in developing cooperative learning systems.

Collaboration among small firms can help achieve economies of scale in human resources development, technology, market research, sales, distribution, and service. For many small companies, the cost of large-scale training or technology development or market planning can be prohibitive unless they can find a way to share those costs with other firms. Collaborative systems or institutions can bring these firms together so that they can do things collectively that they might not be able to afford to do individually. Collaborative arrangements can also help firms discover new ways to pool their individual capabilities to develop, produce and market goods or services that combine their separate expertise. Firms which are individually flexible because of their ability to respond quickly to changing technology or markets can heighten that flexibility through entering into dynamic networks with each other.

But the most important issues here are neither the smallness of the companies nor the economies of scale they can achieve nor the manufacturing networks they can create. Rather, it is the learning which is possible through collaboration. For almost any business seeking to compete in a sophisticated and cluttered international marketplace, and especially at the high end of that marketplace where returns are the greatest, there is just too much to know. Businesses have to develop a kind of collective intelligence. They have to find a way to learn from each other. They can do that through vertical relationships, suppliers learning from their customers and vice versa. And they can do it through horizontal relationships, learning from other firms in the same sector or in the same region.

But these outcomes are possible only if there is a foundation of associative behavior among the firms. However, there are few opportunities for firms to "practice" association and cooperation in America. Especially among smaller firms, entrepreneurship used to be an individualistic activity. As a result, cooperative links among small businesses and between them and their larger customers are developing more slowly in America than almost anywhere else in the industrialized world.

There are several reasons for this relatively low level of consortial behavior in American industry. One certainly is anti-trust legislation, which has had a chilling effect on cooperative behavior among firms. Further, the particular culture of entrepreneurship in America seems to have discouraged firm owners from getting together with each other, at least for the purpose of talking about business. Many small business owners choose to see themselves as "ruggedly individualistic," and act almost as though they were seeking to conform to their stereotype.

Perhaps the sharpest reason for the strikingly different patterns of associative behavior between firms in America and those in other industrialized nations is the relative emphasis on international versus domestic competition. Small and medium-sized manufacturing companies in European and Asian nations have for a long time seen themselves as competing on an international scale. In the U.S., the field of competition has shifted from local to global only very recently for most companies, and some have yet to see or understand that shift. Businesses have seen themselves as fiercely competitive in local markets and have not been inclined to cooperate in international markets. If a firm sees its competition as local, it probably won't establish new arenas of cooperation with similar firms in its own region.

MODERNIZATION AND PUBLIC POLICY

We are increasingly persuaded by our own recent work and the careful research and analysis of others that development strategies and resources should be focused around "groupings" of manufacturing firms, concentrated in regions as "sectors" or "clusters" and linked by similar products, common markets, shared technologies or common resource constraints. We are led to this conclusion by two observations. First, we don't think that one-by-one assistance to individual small firms is either efficient or effective. Second, dramatic changes in the manufacturing environment are shifting the basis of competitive advantage from issues internal to the firm to issues of external relationships among firms.

There are about 380,000 manufacturing establishments in the United States, 95% of them employing fewer than 250 workers. In any particular state or region, even among those with the most highly evolved infrastructure of programs and resources aimed at small manufacturers, it is unlikely that more that 5% have had any substantive engagement with an agency or program that is able to help them. And many of the contacts are just that: brief, sporadic, *ad hoc* contacts focused around a narrowly defined problem and leading to no particular consequence that significantly increases the ability of the firm to compete in high value markets. Manufacturing modernization, gaining the ability to compete in high value markets, is a complex, on-going process of continuous change and improvement. It is not likely to be affected by isolated, and one-time contacts with public sector "helpers," no matter how experienced or well-intentioned.

In survey after survey, small, medium, and large sized manufacturers report that they do not effectively learn from public sector programs. For most of them, these programs are an alphabet soup of acronyms, a welter of "white noise." When asked, manufacturers report that they learn about new markets, technologies, and business practices from the marketplace, from their engagements with customers, suppliers, cooperators, and competitors. (That itself is, of course, problematic since there are so few well-established mechanisms for linking firms into effective learning relationships.)

To be fair, all the evidence isn't in yet. With new federal leadership, there is a whole lot more attention and resources going into industrial extension systems and manufacturing technology deployment and improvement programs. Some of the savings from the defense budgets cuts are being plowed back into manufacturing assistance centers. But we are more than dubious that continued reliance on one-by-one approaches to helping firms will yield significant results. There are too many of them, they don't know how to use government programs, and they need long-term learning systems, not short-term problem solving.

Of even greater import than the question of efficiency is the matter of the changing nature of competitive advantage. At least at the high value end of markets, things are made by groups of companies working together in multiple and dynamic relationships with rapidly growing inter-dependency among firms in these groups or production networks. Increasingly, competitive advantage must be found in the flexibility that can be achieved though flexible external organization.

High value manufacturing is characterized by highly segmented, volatile, international markets demanding high quality and high service. To meet those requirements, producers must deploy advanced technologies operated by highly skilled workers with the most sophisticated management practices. They must learn how to specialize and how to combine their specialized capabilities with others. Inter-dependency grows both vertically (up, with their best customers and back, with their preferred suppliers) and horizontally as they learn to cooperate with others to optimize their mutual complementarity.

It is in these new relationships where learning occurs at an accelerated rate. And at the core, getting or being competitive in high value manufacturing is learning about markets and marketing practices; quality requirements and quality improvement practices; design, engineering, production and distribution technologies; capital and credit management techniques; and evolving skill requirements and new forms of work organization. These are not so much things to learn as they are dimensions of change around which learning must be organized. Internal organization must support learning within the firm. But the real frontier of the "learning organization" lies in the linkages among firms. The stronger and deeper that high value firms are linked to each other, the more and the quicker they can learn. In this sense, manufacturing modernization assistance programs specifically, and economic development programs generally, should be about creating learning systems, not just transferring units of technology or business knowledge.

The relative efficiency and effectiveness of sector or cluster or networking strategies versus one-by-one approaches seems pretty clear. Programs that focus around groups of firms can reach larger numbers; they can promote more industry leadership in the design and implementation of the programs; they can sort out more clearly the long-term factors that will help shape competitiveness; they can ride with, not against, important trends toward "partnering," "networking," and other forms of collaboration among firms; they can promote shared learning among the firms; they can create peer group reference or benchmarking systems that show firms what better practice and best practice looks like; and they can create economies of scale and scope that reduce the unit cost of adapting technology, training workers and installing advanced business practices.

But the assistance strategies of most development agencies persist in viewing the individual firm as the unit of analysis and the target of policy. Is this because the managers of these programs just don't know any better, because they lack information, means and methods to re-configure their policies around sectors or clusters? Or do they have the information but fail to implement these new approaches out of simple inertia? Or is the problem that there are hard political realities at work in development that drive program managers, often against their will and better judgment, to focus on a few trees instead of the larger forests.

There are a lot of votes for inertia as the culprit here. In many ways, the whole "culture" of economic development is about "deal-making" with individual companies. In a sense, development practitioners have just kept on using the same "deal-making" style they employed in the no longer fashionable recruitment model of development. Much of the innovation in development policy has taken place within a fairly narrow range of conventional thinking about what economic development strategy is all about. The programs are almost always far too poorly financed and thinly staffed to have much likelihood of success. They are able to provide assistance to a very small fraction of the businesses that need help. They are fragmented and hard to hold accountable. Even if businesses that do take advantage of some of the service programs are frequently happy to attest to their beneficial results, such a small number of firms receive these benefits that it is difficult to claim any lasting effect. These programs may focus on the right problems, but they are just too small to provide the right solutions and they always will be.

But it isn't fair to attribute these policy shortcomings merely to the inability of development practitioners to shed their deal-making skins in favor of "collectivist" strategies. Those who frame the big policies in the legislative and executive branches of government are a big share of the problem. They tend to see complex economic issues as constituent problems. Those who have had to run development programs in states and regions can attest to the virtual absence of political support for "sectors, networks and clusters" as opposed to individual firms. Shifting development resources away from individual deal-making toward collaborative approaches to gaining competitiveness runs against the political grain.

So how do we change policy if inertia and politics, two very powerful forces, are both against us? Perhaps we appeal to the firms themselves, at least to those who are leading the way toward high value manufacturing through collaboration.

Our better firms are beginning to understand that being competitive is about being good at learning. They are beginning to search for learning systems. Granted, most of them might have a little trouble seeing their governments as learning systems. But they do see their linkages with other firms' customers, suppliers, competitors and collaborators as crucial to their market success. When and as government agencies begin to work with them in building the linkages among firms that will help create these learning systems, the firms will create a constituency.

8

GOVERNMENT RESPONSE TO STRUCTURAL CHANGE AND THE IMPLEMENTATION OF MORE FLEXIBLE MANAGEMENT AND PRODUCTION SYSTEMS

J.W. Wheeler
Hudson Institute, Herman Kahn Center, P.O. Box 26-919,
Indianapolis, IN 46226 USA

INTRODUCTION

For well over a decade, analysts from many disciplines have engaged in a vigorous debate over the nature and consequences of the great changes sweeping the structure, organization, and location of industries and enterprises. This debate has often been couched in the context of a possible emergence of a new era of "flexible accumulation." From a pragmatic perspective, much of this debate has had limited direct policy relevance, but the concepts have served to frame the policy debate and guided policy decisions in the face of inadequate understanding of the changes at work. Several key areas of consensus and lines of analysis have emerged that help guide government actions.[1]

Among the more important of these is the spreading recognition that place must be a crucial policymaking dimension. In practice, of course, local impact has always been a major determinate of specific government actions (on the basis of pure pork barrel politics, if for no other reason), but the policy debate has often neglected any careful spatial analysis. This chapter will discuss a number of the more important issues for government action emerging from the flexible accumulation debate. The choice of issues is highly idiosyncratic. No effort will be made to provide a systematic and balanced review of all policy-relevant implications in the available literature.

BACKGROUND

The flexible accumulation debate is both a variant and an extension of the debate that raged over the emergence of a "post-industrial" society some two decades ago, and it parallels major aspects of the research into the causes and consequences of the growing service economy and into the emergence of an information economy (Kahn, 1979; Bell, 1976). And like these predecessors, basic concepts, definition of terms, and empirical evaluation of theoretical constructs have been major stumbling blocks to analysis. One of the strongest conclusions from the literature is that there is no clear emergence of a new era. Flexible production and management systems are as much a desiderata as they are a reality, and the available evidence points as much to adaptive strategies of enterprises facing great stress as to a new paradigm. Nevertheless, a transition is clearly underway, even if it cannot, as yet, be cleanly described in a clear theoretical construct.

First, firms are clearly seeking to adopt an array of more flexible production practices. In many cases, such efforts are more experimental or partial than fully adopted, but successful firms appear to have implemented a variety of flexibility initiatives. Gertler (1992, p. 260) identifies six types to which I add a seventh:[2]

i. a more flexible use of workers and machines; [including the adoption of "flexible manufacturing in its various forms and variants]

ii. more flexible inter-firm relations . . . ; [creating a complex network of both cooperative and competitive interactions]

iii. more flexible relations with the market, driven by intensified competition and a consequent shortening of the product life cycles . . . ; [new products, new variants, better and faster service and support]

iv. a reduction in the amount of unrealized capital tied up in inventories; [and work in process]

v. changes in social institutions to foster more flexible employment relations; [union contracts, subcontracting, use of contract workers]

vi. the breaking down of previous barriers to the mobility of capital between sectors and between places . . . [the dramatic improvements in and convergence of computing and telecommunications, combined with advances in management processes have permitted radical unbundling of processes]

vii. reduction of internal rigidities through organizational changes, such as flattening the bureaucratic hierarchy, partnering, selling, or closing non-core business activities, and devolution of decision making authority.

Second, competition among firms has become increasingly complex.

i. Depending upon the product and specific market developments, non-price factors (quality, delivery schedules, service, reliability, etc.) vie with price as the dominate dimensions of competitiveness.

ii. Arms-length relationships in a competitive marketplace are increasingly admixed with cooperative arrangements. The most common have been changes in supplier-customer relationships towards longer-term stable collaboration. These vertical arrangements have been accompanied by new horizontal links as firms in competition create joint ventures and, with increasing frequency, strategic alliances to deal with specific challenges (e.g., R&D in response to regulatory change), develop new products which involve high cost or risk, and even to segment and share markets. Complex networking has become a major aspect of firm responses to managing intensified competition.[3]

iii. Rather than being the domain of small and medium-sized firms, such flexible arrangements have spread into the largest of global corporations as they struggle to remain competitive in a rapidly changing market context. Flexible accumulation thus does not portend the demise of large corporations, as some early analysis suggested, but rather implies a changing array of relationships among large firms, and between large and small firms. These changes raise a variety of legal and regulatory concerns that must be dealt with at various jurisdictional levels, if localities and nation-states are going to be good hosts for private economic progress. Such concerns range from the anti-trust implications of these new complex networks, to regulatory and risk exposure aspects of new/potential financial programs, to the constraints on contracting and partnering extant in commercial and tort law, accounting standards, etc., build up over time for very different economic conditions.

Third, as industries and firms adjust to the rapidly changing competitive environment, new technologies and organizational innovations have created a particularly fluid environment for investment and location decisions. As a result, community characteristics and policy actions appear to have a greater impact than before on firm decisions. The flexibility of the policy and institutional framework to accommodate change is therefore vital to both national and local economic vitality.

i. Advances in telecommunications and distributed computing, building upon the long-existing trend of declining transportation costs and transit times, have permitted the unbundling of many business activities into widely dispersed locations. Prominent examples are the transfer of headquarters back office operations out of high cost urban locations, the establishment of 1–800 and other operator-intensive communications services in lower cost communities along major telecommunication corridors, and the transfer of major steps in software coding to such places as India and Egypt. The explosion of rapid delivery services (e.g., Federal Express) both depend upon and serve as critical infrastructure for these global changes.

ii. At the same time, these advances have laid the foundation for a wide array of new products and forms of commercial enterprise that were previously unthinkable, ranging from on-line retailing, to distance learning and remote health care, to the widespread use of electronic mail, and many others. Some analysts argue that a very large share of new products and services introduced in coming years will be based upon the micro-electronics/telecommunications revolution (Keyworth and Abell, 1990, p. 6).

iii. For many industrial activities, scale economies have become less important (with more flexible manufacturing capabilities and the fragmentation of markets), permitting both an increase in subcontracting and partnering, and the geographic dispersal of production stages according to a variety of cost and competitiveness criteria (e.g., labor cost, skill levels and workplace attitudes; taxes, fees and other direct costs of government; regulatory burden; availability, quality, and costs of government and infrastructure services; proximity to key markets or sources of supply; etc.).

iv. In many industries, automated processes have greatly reduced the number of employees required for any given level of production, while also raising average worker education and skill requirements. This both expands the number of potential sites and creates a bias towards more affluent suburban and ex-urban locations.[4]

v. The growth of outsourcing by large companies has created a number of highly competitive markets for intermediate products and technical and business services. In turn, as larger companies have become increasingly dependent on outsourcing, their investment and location decisions are more strongly influenced by ready access to suppliers and subcontractors.

Do all of these changes reflect an emergence of flexible accumulation as a new paradigm of economic activity? Or, as this author is inclined to believe, is it more fruitful to see flexible accumulation as an important new dimension of the evolutionary adjustment of market economies to changing conditions, a dimension that presents unique policy challenges and opportunities to various levels of governance?[5]

THE POLICY CONTEXT

Policymakers are operating in an environment characterized by a number of overarching forces. The challenge is, of course, to design policies that minimize the cost and/or take advantage of these longer-term forces while also responding to short-term political pressures, that may or may not be consistent with longer-term local or national economic interests. Some of these overarching forces are noted briefly below.

Global and International Forces

The globalization of economic activity continues unabated. With the exception of global reaction to the first oil-shock, international trade has grown faster than domestic trade every year since the 1940s (GATT, 1994). Although more volatile than trade, international investment flows (portfolio or direct) have grown faster than international trade. In the advanced industrial countries, and in much of the developing world, it is accurate to say that no industry is immune from international commerce. Looking forward, barring a catastrophic collapse of the global economy, all trends point to a continuation of this trend.

The recently signed Uruguay Round agreement reduces tariffs of goods by about one-third, extends market liberalization into areas long outside the GATT mandate (services, agriculture, and textiles and apparel), and puts into place a large number of other arrangements (on intellectual property, subsidies, standards, and trade related investment measures, to mention only the most important). Once fully implemented, it will stimulate a tremendous increase in cross-border trade and investment. In addition to global liberalization, regional groups of states are dropping barriers even more rapidly, most notably among members of the European Union and the North American Free Trade Agreement. Many developing states and former members of Soviet Union and the Warsaw Pact are unilaterally reducing barriers to international commerce in an effort to stimulate domestic progress. Indeed, we may be seeing an unprecedented degree of economic liberalization worldwide.

Although much debate exists over details, the basic lessons of economic progress are available to any state within which the political desire and will for implementation can be mobilized. The newly industrializing economies and their immediate successors exemplify what can be done. Perhaps even more impressive is the success of the partial reforms in China. Efforts to pursue variations of capitalist development now exist virtually worldwide. Not all of these efforts will succeed, but the direction of change is towards larger and more open markets and the emergence of greater numbers of competitors across the industrial spectrum.

Firms ready and able to compete in this environment, the communities within which they reside, and most nation-states will benefit greatly from these trends. Competition, however, will intensify and industries, firms, and communities already under stress due to competitive adjustments can expect even greater market pressures. The implied rapid economic changes will force elected officials and policymakers to respond, whether reactively or proactively.

National Forces

National forces vary considerably from country to country. Several crosscutting issues are best assessed as national rather than global due to the highly differential fashion in which states have responded.

Global diffusion of technology and access to higher education have created a dynamic in which each state's policies and its business and workplace culture profoundly influence economic progress. Unless firms in a state practice or benchmark themselves against best commercial practice in their worldwide industry, and develop means to follow what appears to be continued rapid technological change, they inevitably face negative competitive developments. Not surprisingly, technology policies and, perhaps even more importantly, policies that influence the insertion of technology into domestic products and processes (often called technology deployment) have become a major area of domestic policy debate and a frontier of dispute among nations. Certainly it has been a key issue in all major recent international trade agreements and negotiations. Location and ownership of investments in technology have become an increasingly important aspect of the policy debate.[6]

A second crosscutting theme has been that of employment creation. The close historical link between economic growth and employment generation has been decoupled. Despite an array of training, adjustment, and clearinghouse programs among the advanced countries, employment policy has been primarily macroeconomic in nature, backed up by various welfare state safety nets. Not only have these policies failed to correct the high sustained rates of unemployment in western Europe, but more appropriate microeconomic policies targeted at reducing the deep seated rigidities of European markets have proven difficult to implement. For example, the French government was forced to ignominiously back down from an interesting plan to flexibly reduce youth minimum wages in a effort to not only expand youth employment opportunities, but to help instill good work habits and a longer-term affiliation with the workplace. European Union members generally recognize the imperative to make their labor markets more flexible, and, ever-so-slowly, are making some headway.

American labor markets are far less rigid, although recent policy initiatives have included a bias against employment growth, whatever their merits on other grounds (family leave, efforts to raise the minimum wage, commitment to Davis-Bacon, various health care reform proposals, etc.). Indeed, the U.S. marketplace has created jobs at an awesome rate over the past 12–15 years. Aggregate employment problems in the U.S. are more macroeconomic than in Europe, while the compositional problems are much more race and location concentrated. Employment creation in the U.S. must deal with the concentration of the unemployed in central urban districts, primarily in old industrial areas, where minorities make up the largest portion of the population. Some of the location specific issues will be dealt with below.

A related issue is dealing with the mismatch between the demand for skills implied by the changing industrial structure and that likely to be available based on spatial characteristics of demographic projections. This is perhaps most serious in the U.S., but has important European dimensions based on social, class, and locational rigidities.

A third crosscutting factor is how to build and maintain national competitiveness, however defined. After years of industrial restructuring and with the help of a favorable exchange rate, many U.S. industries are enjoying greatly improved export competitiveness, though sales are temporarily depressed due to stagnant overseas market growth, and the trade and current account balances will remain in deficit as long as domestic absorption (spending on goods and services) exceeds domestic production. Such export vitality is a sign of flexibility and resiliency.

A great many West European companies have maintained their competitive edge through extensive capitalization, international diversification or in some cases with help from their governments. However, the limits of such strategies have increasingly run up against their high domestic cost structures. Implementation of market-deepening and regulatory harmonization policies will buy some important time as firms' restructuring for a larger market produce significant Europe-wide efficiency gains. However, intra-EU stresses resulting from these adjustments will, at the same time, intensify the decline of old industrial areas.

As can be seen in policy difficulties encountered in dealing with the nation's financial crises and the cascading effects of the strong yen, Japan is just beginning to face the adjustment pressures of becoming a mature economy, though from a fortunate posture of containing few industries in great distress.

A fourth crosscutting theme is dealing with an aging population. All of the advanced industrial countries at some point must deal with a substantial increase in the proportion of aged, with all that implies for healthcare costs, social security systems, and economic dynamism. Aging has many other crosscutting impacts besides social welfare program budgets. For example, on the positive side, a growth in the proportion of the middle-aged implies growth in average productivity and savings rates. On the negative side, population aging also implies decreasing mobility and flexibility of the labor force. Little attention has been paid to sorting out the important interactions among demographic change and the imperatives for flexible accumulation.

Regional and Local Forces

It is at the regional and local levels where the metaphorical rubber meets the road. Firms open or close, hire or fire, and expand or contract at the local level. One of the great tensions, at least in the U.S., has been that political jurisdiction often has little relation to that geographical area in which externalities and agglomeration effects tend to operate. In part, such tension contributed to the widespread city annexation drives during the 1950–60s, the formation of regional authorities (the New York Port Authority, various integrated districts across political lines to deal with a special activity such as water and sewage), unification of independent political jurisdictions (such as the formation of the Indianapolis-Marion County UNIGOV in Indiana), and more recently the formation of regional economic development alliances of one sort or another.

Indeed, the popularity of regional economic development strategies is based on the assessment that a coherent economic region is generally larger than most local political jurisdictions. Thus, all gain by cooperating rather than competing, pooling resources in ways to get high leverage on firm decision making, and reducing redundant programs and staffs. Inevitably, however, such a strategy has not dealt effectively with pockets of distress within the larger area. Firms may locate or expand jobs and investments in the broader area, but rarely in the core and with few direct jobs for the most disadvantaged.

Only a limited number of large new investments are looking for a site at any given time. Local jurisdictions have tended to spend a inordinate share of their economic development resources in competition, often giving what are later seen as excessive incentives. In recent years, many jurisdictions have backed away from such excess competition, putting incentive packages together that are based on more carefully calculated costs and benefits, as well as taking a much harder look at the direct and indirect impacts of different forms of new economic activity.

In many areas of the U.S., attention and resources have shifted toward growing from within, creating new activities building upon local assets, or diversifying the economic base rather than a shotgun approach to investment attraction. Some of these areas have been so successful as to lead to a designation as an entrepreneurial city. Many of the characteristics of an entrepreneurial city reflect the general description of flexibility translated into the public sector: trust-based flexible alliances and partnerships among participants that otherwise both cooperate and compete, concentrated on investment in activities that build on and extend agglomeration economies which draw mainly from a local labor market.

Another major local concern, though distributed very unevenly across the U.S., is the consequences of dramatic defense cuts on top of the ongoing restructuring of the private sector. Some of the success stories of flexible manufacturing based on high technology in part were supported by responding as subcontractors to a core military demand that tended to be highly insensitive to cost. This simply will not continue into the future.

Political competition, power relationships, and the dynamics of leadership and elite interaction all affect the ability of a jurisdiction to accommodate the requests for increased institutional flexibility from firms undergoing structural transformation, especially in the absence of a careful economic assessment of the costs and benefits of offering such flexibility. They also affect the ability to develop a strategy of development that relatively benefits one subset of community interests.

SPECIFIC POLICY CHALLENGES

In the face of these diverse forces affecting economic conditions and the policy process, the various dimensions of flexible accumulation present a series of reasonably well-defined policy challenges as well as raise several key questions.

Regardless of the precise definition of flexible accumulation used, intensification of market competition and continued rapid technological change create a strong demand for government policies that enhance firm flexibility, whether through its relations with labor, its response to regulatory mandates, or its ability to collaborate with competitors for specific purposes. Those nation-states and jurisdictions within nation-states that can best respond to these demands without serious cost to other political and social goals are likely to be attractive locations for firms shifting towards more flexible production and organizational structures. Although regulatory relief and other forms of flexibility have served as an early element of many policy initiatives, only rarely has such relief actually been implemented, especially at the federal level. Some experimentation has occurred in environmental regulation, such as with the bubble approach to emissions control. Slightly more success is seen at the regional and local levels, as some cities and states are slowly beginning to set up processes to reduce regulatory delays and cost. Progress is hampered by the explosion of federal regulation and the lack of integration and cooperation among federal, state, and local processes to solve these problems.

In this regard, simply introducing new machines and management structures does not assure successful adoption of flexible practices (the range of success and failure in implementing Total Quality Management in similar settings within and among firms offers important lessons). The goal is to inspire both workers and managers to join together in a commitment to success of the new or restructured business venture, and this requires building trust, sharing information, and demonstrating a long-term commitment to cooperation. Particularly at the local level, there may be important intermediary and bully pulpit roles for government and community leaders to help alter the local mindset and prepare managers and workers for new roles and responsibilities. This can be particularly effective in an urban setting in which similar practices are being adopted in the public sector (e.g., an entrepreneurial city).

A similar, but more operational role can be defined for a government seeking to facilitate the kind of linkages among firms upon which an industrial or educational network can be built. This can range from serving as a catalyst for bringing potential network partners together to discuss a particular opportunity or issue, to taking on an active broker role in putting a deal together (to which the government may contribute resources or offer certain policy commitments), to becoming a full partner in the network with clearly defined roles and operational responsibilities. In any of these roles, the government can by its presence help legitimize the activity and directly or indirectly offer protections to potential network partners who see possible business or legal risks resulting from their participation. It can also help build the trust among participants that such networks depend upon. The more heterogeneous the community, the more important such a government role may be. With the continued pressures on large companies to concentrate on their core competencies, a major state and community policy role is to find locally viable ways to help induce the formation of networks of subcontractors that large firms are increasingly dependent upon. For example, identification of key gaps in business services or local industrial support activities, and why such gaps exist, can often help focus policy or regulatory initiatives.

As noted above, the adoption of flexible organizational and production methods accentuates the mismatch between skills needed by business and those held by the existing labor force and future new entrants. A major source of the problems in education and training is poor policy, and it is incumbent upon various levels of government to reverse these errors and take what actions are possible to help deal with the mal-educated. In itself, this is an area of great political confrontation. Many experiments are underway, and some important lessons are already well documented, such as the high degree of success from on-the-job-training programs. Glitzy new high-tech programs often distract from getting the basics right. Workforce preparation will be among the most serious challenges to success in every community, as they adapt to the increasing imperative for local flexibility.

One of the core aspects of many discussions of flexible accumulation is a reliance on the use or introduction of innovations in micro-electronics and telecommunications. These are core drivers of modern technological and industrial progress, and are fundamental to the restructuring of economic activity. The U.S. remains a leader in technological innovation. Yet it also continues to show a distressing lag in technology deployment. Experiences reported from a variety of economic development programs suggest the great potential inherent in helping to facilitate the introduction of micro-electronics and information technology into firms focused on more traditional service and manufacturing processes.[7]

Research has demonstrated the importance of large firms in supporting many flexible networks, by serving as the market for their output, and by providing capital, as well as product and process innovation. Particularly in efforts to link firms from distressed areas into networks, government could well take on the role of helping individual small firms meet acceptable standards of quality and delivery for large firms. Incubators, economic extension services, and so-called manufacturing technology centers, among others, sometimes take on such efforts, but rarely as a primary objective. For smaller firms and firms in traditional industries facing adjustment stress, government can help facilitate information exchange about successful adjustment. This is a traditional government role that, in general, has not been very effectively implemented. The recent experiments with manufacturing extension services around the country is a response to this long-term policy failure.

Further, it is possible to use the government's power-of-the-purse to induce firms to meet certain quality, delivery, and other standards. If the various components of government, as major customers for many goods and services, do not require world class standards, the suppliers have less incentive to invest in achieving those standards and thus are helped to remain less competitive. This is particularly important in mission critical areas where the government is among the major customers or has a major role in setting product standards. Good examples include defense, healthcare, NASA, telecommunications at the national level, but state and local governments also have many avenues to influence local supplier quality through the power of the purse. It is also possible for local governments to help establish small neighborhood companies and networks designed to capture a larger share of the spending power available in distressed areas.[8] Indeed, within the process of privatizing certain government operations, it is possible to implant a number of well developed supplier-customer networks.

CONCLUSIONS

From a policy perspective, the transition to flexibility appears to be best understood as an important new dimension of the complex process of adjustment to rapid technological change and intense competitive pressures. It is not just a technological process, nor is it simply a restructuring of management practices. It must involve a redefinition of the relations, roles and responsibilities between managers and workers. New forms of cooperation must emerge between suppliers and customers, and sometimes among competitors. At the same time, government can be instrumental in laying the foundation for a more adaptive, collaborative policy framework that supports flexibility and change while protecting important social and political goals.

Flexible methods and approaches are increasingly the key to success in today's highly dynamic and competitive marketplace, but not just for firms. It is critical that governments also re-engineer themselves to be more flexible, and utilize more performance and market-based metrics of success for service delivery.

Whatever the economic development strategy, it is critical that governments avoid the mistake of swimming against the tide. Protecting firms from the inevitability of change, absent a transitional strategy of adjustment, is doomed to failure and usually entails much higher social and taxpayer cost.

ENDNOTES

[1]A vast literature addresses different aspects of this debate. A good review of some of the geographic writings can be found in Gertler (1992).

[2]The background section of this chapter draws heavily on Gertler's (1992) analysis.

[3]The automobile industry presents an excellent case study of the highly varied new relations among firms. It is important to note that, with a few exceptions, the cooperative relations among major automotive rivals lay outside what individual companies see as their core competencies.

[4]For a good analysis of the growing mismatch between job requirements and the skill mix of existing and future workers, see Johnston and Packer (1987).

[5]For an excellent discussion of the forces transforming manufacturing today, as well as identification of some useful indicators to judge success, see Vollmann (1994).

[6]See Porter (1990), Costello and Ernst (1992), and a series of studies and projects sponsored by the Competitiveness Council.

[7]Based on discussions with individuals from different states involved with establishing manufacturing extension programs, or in other business support networks, including the U.S. Small Business Administration, at various times during spring/summer 1994.

[8]Examination of income and spending patterns in distressed neighborhoods reveals extensive outflows of purchasing power on products and services that could easily be offered by small local enterprises.

REFERENCES

Bell, D. (1976). *The Coming of Post-Industrial Society*. New York: Basic Books, Inc..

Costello, R. and Ernst, M. (1992). *Regaining U.S. Manufacturing Leadership*. Indianapolis: Hudson Institute.

"General Agreement on Tariffs and Trade," (1994). *International Trade 1993*.

Gertler, M.S. (1992). "Flexibility Revisited: Districts, Nation-States, and the Forces of Production," *Transactions of the Institute of British Geographers NS* 17, (3), pp. 259–78.

Johnston, W.B. and Packer, A.H. (1987). *Workforce 2000*. Indianapolis: Hudson Institute.

Kahn, H. (1979). *World Economic Development*. New York: Marrow.

Keyworth, G.A. II and Abell, B. (1990). *Competitiveness & Telecommunications*. Indianapolis: Hudson Institute, p. 6.

Porter, M.E. (1990). *The Competitive Advantage of Nations*. New York: Free Press.

Vollmann, T.E. (1994). "Transforming the Manufacturing Enterprise: A Key Issue for Competitiveness Today," *World Competitiveness Report 1994*, September, IMD and The World Economic Forum: Lausanne and Geneva, Switzerland, pp. 222–26.

9

FIXED STRUCTURES IN TRANSITION: THE CHANGING DEMAND FOR OFFICE AND INDUSTRIAL INFRASTRUCTURE

Nancey Green Leigh
Department of City Planning, Georgia Institute of Technology, Atlanta, GA 30332-0155 USA

INTRODUCTION

This chapter examines some aspects of change in the demand for infrastructure in the U.S. space economy, relating this to "flexible accumulation," a label given in recent years to the evolving capitalist system. Derivatives of the flexible accumulation label such as flexible specialization and flexible production have subsequently been coined to describe changes in industrial organization accompanying capitalism's evolution. While the labels were first applied to manufacturing reorganization, their perceived relevance for describing reorganization trends among the service sector has been increasing.

From the outset, however, as illustrated in Harvey's sweeping and influential definition of flexible accumulation, infrastructure has not figured heavily in the "flexible" literature:

> Flexible accumulation, as I shall tentatively call it, is marked by a direct confrontation with the rigidities of Fordism. It rests on flexibility with respect to labour processes, labour markets, products and patterns of consumption. It is characterized by the

> emergence of entirely new sectors of production, new ways of providing financial services, new markets, and, above all, greatly intensified rates of commercial, technological, and organizational innovation. It has entrained rapid shifts in the patterning of uneven development, both between sectors and between geographical regions, giving rise, for example, to a vast surge in so-called 'service-sector' employment as well as to entirely new industrial ensembles in hitherto underdeveloped regions—It has also entailed a new round of what I shall call 'time-space compression' . . . in the capitalist world—the time horizons of both private and public decision-making have shrunk, while satellite communication and declining transport costs have made it increasingly possible to spread those decisions immediately over an ever wider and variegated space (Harvey, 1989, p. 147).

Harvey and those generally writing on the subject have focused on "labour processes, labour markets, products, and patterns of consumption." Yet changes in the organization of employment, production, and consumption must inevitably translate into changes in the demand for the structures which house these activities. In the relatively brief time since it was introduced, the term "flexible" has become a buzz word in economic development research literature as well as mainstream business journals. It has also been subject to considerable criticism. As Sayer and Walker have observed:

> the "discourse of flexibility" involves ideological slippage between description, prediction, and prescription which masks the vital political issue of the different interests at stake. . . . (Sayer and Walker, 1992, p. 198).

There are, of course, differences in the degree to which different political interests are place-bound. Greater global integration of production and advances in telecommunications have lessened the attachment of the architects of the new flexible patterns, those Reich (1992, p. 178) calls the "symbolic analysts," to the urban economies in which the exclusive residential communities where they typically reside are located.

We cannot fully understand the evolution of the spatial economy and, in particular, its impacts on economic opportunities for different demographic groups if we do not also consider how the demands for existing patterns of infrastructure, such as industrial and office plant, are shifting. Change in the pattern of location of office and industrial plant also alters the accessibility of service and industrial activities to different labor groups. It may also shift some areas of the metropolitan and regional economy and their residents into decline, and advance the consumption of greenfield, or, previously undeveloped land.

"Fordism" is the label given to what has been claimed to be the previously dominant mode of industrial organization. Some of the characteristics that have been associated with it include assembly-line mass production (introduced by Henry Ford to auto-making), de-skilling of labor, and vertical integration of production. Fordism is associated with the production of large quantities of standardized products using rigid (e.g., inflexible) processes. The scale and organization of these production processes impeded firms' abilities to respond to the ups and downs of the business cycle. Critical of the Fordism/Flexible dichotomy, Sayer and Walker observed that:

> ... capitalist industry has always combined flexibilities and inflexiblities. What may be emerging now are new permutations of each rather than a simple trend towards greater flexibility alone. The variety of such combinations cannot be grasped by inflexible dualistic frameworks which counterpose the old as inflexible to the new as flexible. Indeed, mass production and flexible production need not even be considered as alternatives. What is therefore needed is not an obsessive focus on flexibility but a broader awareness of new forms of the division of labor and new ways of organizing those divisions (Sayer and Walker, 1992, p. 199).

According to Sayer and Walker, it is necessary to distinguish between product flexibility which may not require capital changes within the firm, and flexibility in restructuring. In turn, individual firm concepts must be distinguished from flexible organizational forms, that is, changes in networks or industries (Sayer and Walker, 1992, p. 199). These distinctions will also, of course, influence the changing requirements of within-firm and external infrastructure for offices and industries: the quantity and configuration of space needs, the need for proximity of suppliers in the production process, the need for centralized locations within the urban region, and so on. Determining what these requirements are is a large task and only small pieces are addressed in this chapter.

The remainder of this chapter is divided into three major sections. Each has, within this chapter's overall theme of considering fixed structure in transition, an underlying theme of determining what this means for the fates of the nation's central cities. The oldest of our industrial plants are, of course, located in central cities and their first-ring suburbs. The central business district was the primary site for office buildings through the first eight decades of this century.

The first section focuses on the fixed structures in transition in one of the nation's oldest industrial centers: Milwaukee, Wisconsin. The section contains an account of two examples of what I have called elsewhere "the making and unmaking of industrial wasteland" to demonstrate how new life has been breathed into two obsolete industrial plants, one located in the central city and the other in the earliest of Milwaukee's suburbs (Leigh, 1993). Second, current demands for office and industrial space are reviewed, focusing on locational preferences as well as changing technologies, to gain insight into how fixed structures are being utilized in an age of increased flexibility and mobility of capitalist industry. Third, the section focuses on a substantial barrier—environmental contamination and federal regulations—to reusing older office and industrial buildings in central cities which can deter efficient use of existing infrastructure, and significantly dim prospects for revitalizing central cities.

MILWAUKEE: A TRADITIONAL INDUSTRIAL CITY

Milwaukee, Wisconsin experienced massive deindustrialization between 1979 and 1983, and in its subsequent reindustrialization phase, its regional economy, like the national economy, saw a significant shift of jobs to services. Also parallel to national trends, these growing service industries, as well as many of the remaining manufacturing industries, have been sited primarily in areas outside of the central city, thereby significantly altering the region's industrial location pattern.

Two exceptions to the siting of services outside the traditional city region, however, have led to redevelopment of sites where two of Milwaukee's most significant manufacturers died. These two producers were Schlitz Brewery and Allis-Chalmers (A-C). In the 1910 Census, Milwaukee ranked first among the top ten manufacturing cities in the country. By this time, Schlitz Brewery had been built alongside the Milwaukee River next to tanneries and other breweries and Allis-Chalmers had built its production site in West Allis, which became a first-ring suburb of Milwaukee. Many major Milwaukee firms had the hallmarks of Fordist industries: vertical integration, mass production, and stable oligopolistic market structures (Storper, 1989, pp. 273–75). Allis-Chalmers was the big daddy of them all, employing 17,000 workers at its West Allis location in its peak.

Between 1983 and 1993, the old Schlitz industrial site underwent a total transformation from a crowded, aging industrial facility to a sleekly landscaped office park, featuring restored and externally charming turn-of-the-century factory buildings. The demise of the Allis-Chalmers Corporation left gaping holes in the industrial landscape of West Allis that are gradually being filled. Before the end of this decade, the A-C site, which is over four times larger than the Schlitz site, should also have completed its transformation. The chronicles of Schlitz and Allis-Chalmers demise and their successors provides insight into the utility of what is often thought to be "industrial wasteland" for the evolving urban economy, and specifically, for what could be characterized as one of the nation's centers of "High Fordism."[1]

The Schlitz Brewery

Schlitz Brewery was begun in 1849.[2] Its well-known motto, "The Beer that made Milwaukee Famous," originated in 1871, when the company sent emergency beer shipments to Chicago to alleviate water shortages that occurred at the time of that city's Great Fire. The company's original production site, located along the Milwaukee River, was built of the "Cream City" off-white brick produced in the region. Built in the late 1890s and early 1900s, the buildings of the brewery had names like Brewhouse A and B, Bottlehouse A and B, the Malthouse, and the Stables. The quality and beauty in the details of these buildings (such as the ornate iron railings with carvings of wheat sheaves) are representative of the skilled European craftsmanship of the era and have stood the test of time.

The Schlitz Brewery was important to Milwaukee as a source of employment and identity for most of the 20th century. As a producer within the company's national network, the Milwaukee site's peak years occurred in the 1950s when its modern Bottlehouse C was considered state-of the art. At that time it produced more beer than any other brewery in the world. Less than 30 years later, however, the facility was considered obsolete.

During the 1970s, Schlitz was subject to legal actions alleging various forms of unfair trade including price fixing, monopoly violations, trademark infringement, improper marketing practices, and improper maintenance of records relating to such practices. It also experienced declining sales. These forces put the company on a downward spiral from which it did not recover.

The Milwaukee Brewery was sold to Strohs (another national beer-maker) which did not bring it back into operation. Instead, Strohs ended up selling the site to local developers for $4.7 million in 1983 (Grunau, 1991). The purchase included 47 acres that encompassed approximately five to six blocks, 2.5 million square feet of building, and more than 2,000 feet of frontage on the Milwaukee River.

At the time of the sale, the site contained many obsolete structures in a larger area made up of other obsolete structures from Milwaukee's early industrial days. When attempts to generate new industrial uses at the site were unsuccessful, the developers opted to develop the Schlitz Brewery as an office park based on the rationale that they could provide office space and parking in close proximity to downtown Milwaukee and near a labor pool for costs of $13.50 to $15.00 per square foot (including parking and utilities development) compared to low $20s costs in the downtown. These development costs were comparable to, or even lower than, those of the area's suburban parks. In the process of converting the industrial site to an office park, the developers tore down one million square feet of buildings on the site to create a "suburban" parking ratio of 4 spaces per 1,000 square feet.

By 1992, Schlitz Park had around 4,000 employees, the profile of which, according to the developers, mirrored a Milwaukee population that is 30% minority. In addition, 75% of the Park's labor force is female. The employees have come primarily from the North and East sides and the Inner City of Milwaukee. The 4,000 jobs at the Park are not the unionized manufacturing-type jobs that the former Schlitz Brewery provided, with high wages and benefit levels. The majority of the jobs are in services and trade. The Finance, Insurance, and Real Estate sector employs well over 2,000 at the Park and these positions are predominantly with the insurance firm Blue Cross-Blue Shield.

A great irony of the Schlitz Brewery's transformation to an office park is that its once state-of-the art bottling facility in Bottlehouse C has now become a service factory floor for claims processors of Blue Cross-Blue Shield. The main production floor of the Bottlehouse has been carpeted and painted. Freestanding partitions are arranged maze-like along the entire span of the floor to provide work cubicles for the predominantly female labor force occupying the vast room.

Allis-Chalmers

The Allis-Chalmers corporation was extraordinarily diversified and engaged in the production of nearly all of the significant technologies which have driven cycles of industrial transformation in this century. Its trademark was "the Company of the Four Powers—Steam, Gas, Water, and Electricity" (Wendel, 1988, p. 4). The 100-acre site it acquired in 1902 became known as "a city within a city." It contained 21 miles of railroad tracks and five miles of roadways, had almost 200 acres of industrial buildings, and at its peak employed as many as 17,000 workers. In addition, the Allis-Chalmers site had its own powerhouse, clubhouse, fire department, and hospital (Wendel, 1988).

Allis-Chalmers was a $2.5 billion corporation when it filed for bankruptcy in 1987. Created out of the bankruptcy court, the Allis-Chalmers Reorganization Trust was assigned responsibility for disposal of noncash assets, including the properties on the West Allis site. In its subsequent transformation, the Allis-Chalmers site has become a striking microcosm of the nation's industrial transformation. Now located on the site are a foreign-owned manufacturing facility (a spinoff of A-C), a corporate headquarters, a technical college branch, a T.V. studio, and a host to the nation's well-known discount retailers (including the Walmart-affiliate, Sam's Wholesale Club). The employment on the site ranges from top corporate management to "good" unionized manufacturing jobs to low-pay, part-time retail jobs. In this transformation, the bifurcation in the earnings distribution of the employees on the site appears to have followed that of recent national trends. As such, West Allis suffers from the loss of good-paying jobs that enabled blue collar workers in the postwar period to have middle class lifestyles. On the other hand, the demise of the Allis-Chalmers corporation has led a diversification of the community's economic base so that no future business failure could be as devastating.

Redevelopment Implications

The redevelopment of Schlitz Brewery and the Allis-Chalmers sites yielded positive benefits for local public finance. The development financing for the Schlitz Park was provided by the City of Milwaukee through two Tax Incremental Finance (TIF) districts. These TIFs have been quite successful. The Schlitz Park's property value was assessed at $4.1 million in 1986, and by 1991, this figure had increased to $35.7 million. Recent property taxes have risen to $1.3 million from $200,000 at the time the site was purchased from Strohs (Kass, 1992). The Park's development has helped in the revitalization of the surrounding area, and specifically the residential area called Brewers' Hill. Thus, increased property taxes through housing rehabilitation and rising prices in that area are spin-offs from the Schlitz Park development. In the case of Allis-Chalmers, the development of new uses on the site increased property taxes from slightly more than $15 million in 1983, to a little over $30 million in 1991.

The Schlitz and Allis-Chalmers redevelopments provide an enlightening demonstration of how abandoned industrial sites in older central cities can be used to compete with suburban office park developments. The objectives driving the movement of "back office" production from the city center to cheaper suburban sites over the past decade can be achieved within the central city (Nelson, 1986). Abandoned industrial land can be obtained and redeveloped relatively inexpensively. Employers can tap into the large pool of lower-wage female labor within the central city. As the suburban labor force shows signs of being tapped out, suburban office parks increasingly have been importing central city labor to meet their growing employment needs. Furthermore, lack of efficient public transportation and the unreliability of low-wage labor's private transportation have created problems for suburban employers who have been importing labor from the inner city. A centrally located back office park like Schlitz which is on the edge of the inner city does not face these barriers.

The two Milwaukee examples presented in this section are illustrative of shifts from industrial to office use.The next section looks at recent data on the demand for office and industrial space as well as some of the technology and organization practices that may significantly alter this pattern of demand.

OFFICE INFRASTRUCTURE

The decade of the 'eighties left both the office and industrial infrastructures of central cities in rather sorry states. By now the overbuilding in the office market of major cities is widely recognized. This trend was stimulated by the 1981 Recovery Act which encouraged investment that did not yield positive returns before taxes. Further, this trend is also associated with the deregulation of the Savings & Loan (S&L) industry, and its subsequent crisis. More than one observer has likened the attitude of the office construction market characterizing the 1980s as akin to the "build it and they will come" message of the popular film *Field of Dreams*. In reality, the contractors built, but new occupants did not arrive *en masse*. Instead, occupancy rates declined in general, and there was a trend toward vacating older Class B and C office buildings for new Class A structures, either in the central business district or in suburban office parks. The financing of new office market development came to a halt as a consequence of The Financial Institutions Reform, Recovery and Enforcement Act of 1989. The new capital mandates imposed on S&Ls, as well as the banking industry, effectively eliminated commercial real estate lending. Older office buildings in central cities have not particularly benefitted from the halt of office construction, however. First, there is still a considerable amount of new inventory to be absorbed from the eighties building binge. Second, recent federal legislation adversely affects the desirability and costs associated with acquiring and maintaining older structures. The 1992 Americans with Disabilities Act can require costly retrofitting of older structures. The desirability of property constructed before 1970, when the use of asbestos-containing material was widespread, is negatively impacted by 1990 asbestos abatement legislation as well as EPA remediation rules (Nunnink, 1994, p. 2).

At present, around 46% of total U.S. office inventory is found in central business districts, and the vacancy rate has been estimated to be 19.1% (Nunnink, 1994, p. 16). Many of these districts contain decaying, obsolete buildings that are affected by the legislation cited above. Consequently, it is expected that many of these buildings will be abandoned or retrofitted to other uses. Further, "The distance between Class 'A' and other classes will continue to widen as tenants leave older properties to occupy post-1980s constructed buildings and will provide positive growth to an already stronger suburban market" (Nunnink, 1994, p. 16).

Table 9–1 provides vacancy rates and change in total value of office property by central business district and suburban location. These data represent an average of data that have been collected for 41 U.S. metropolitan areas. We can see from the data that the demand for suburban office buildings is considerably stronger than that for central business district property: vacancy rates are lower, the rate of absorption of space is nearly four times greater, and the rate of devaluation of property has been greater in the CBD market than in the suburban market

Traditionally, growth in the demand for office space has been linked to growth in the Finance, Insurance, and Real Estate, and business service sectors. However, the correlation between employment growth in these sectors and the demand for office space and related infrastructure is likely to lessen in the future because of technological changes occurring in the office workplace. Telecommuting and electronic data storage

Table 9–1: The State of CBD and Suburban Office Markets, 1993.

Location	*Inventory (Sq.Ft.)*	*Vacancy Rate (%)*	*Vacancy (Sq.Ft.)*	*Avg. Ann. Net Absorp. 1990–1993[1] (Sq.Ft.)*	*Total Value Change 1990–1993[2] (%)*
Central Business District	1,152,675,706	19.10	220,236,350	6,428,906	–22.70
Suburban	1,341,563,368	17.90	239,479,190	23,289,330	–17.60

Notes:
[1]Average of annual totals of 41 U.S. metropolitan market areas.
[2]Sum of annual averages of 41 U.S. metropolitan market areas.

Source: Derived from Nunnink, K.K. (Ed.). (1994). *Viewpoint 1994*, Minneapolis: Valuation International, Ltd., February.

are resulting in decreased work and file storage space requirements overall. Information technologies are also facilitating the movement toward "hoteling," "virtual offices," and "virtual corporations," all of which may decrease firms' space requirements. An alternative way to describe the effect of this trend is to project an increase in employee-to-floor area ratios. "Hoteling" is the label given to corporate trends away from providing individual offices or workspaces to employees whose jobs require them to spend much of their time out of the office. Instead, such employees either share desks with other employees on a permanent basis, or, there is office space that is never permanently assigned to any one employee but, rather, is "checked out" to employees on days in which they will be at the corporate office. AT&T began a hoteling program in 1991. It sees the potential to reduce its annual costs for facilities by hundreds of millions of dollars (Bell, 1993). Ernst & Young, one of the Big Six accounting firms expects to save $40 million annually from reduced occupancy costs associated with its hoteling program, while another Big Six firm, Arthur Andersen, expects to save $25 million annually from its hoteling program (Bell, 1993). Hoteling employees may also spend significant amounts of their time in the offices of clients where they are assigned temporary space.

The term "virtual office" has been coined to describe the transitory work environment of employees without permanent office assignments. In parallel, the term "virtual corporation" has been coined by Jan Hopeland of the Digital Corporation "to describe a temporary alliance of staffers from independent companies uniting to exploit a specific opportunity. . ." (Ogilvie, 1994, p. 30).

Hoteling and virtual corporation trends appear to represent a parallel of the search for increased flexibility in the organization of services production that we have observed for industrial production. Vertical disintegration and specialized subcontracting have been basic trends characterizing what some have labeled the demise of Fordism. Large service-producing firms now appear to be decomposing to reduce fixed plant/office requirements. In fact, the firm of Arthur Andersen even calls its office hoteling program "Just-In-Time Offices" (Bell, 1993, p. 1298). Employees in services and research are expected to become increasingly specialized "as they fine-tune their expertise via projects with different companies" (Ogilvie, 1994, p. 30).

The trends in the office market suggest further impetus for suburbanization of work and home, thereby reducing the demand for in-place infrastructure of the central city. To add to the trends discussed, corporations are increasingly inviting employees to work from home and use non-territorial offices for team or client meetings. The overwhelming residential preference is, of course, for a suburban location, as the 1990 Census reported for the first time nearly half of the nation's population resided in the suburbs. Furthermore, predictions of the emergence of "fully integrated office/ residential communes [that turn] many people's commute into an elevator ride or a walk across the hall" suggest the possibility of a very different spatial pattern of office infrastructure (Bell, 1993, p. 1298).

INDUSTRIAL INFRASTRUCTURE

The state of the national industrial market could, on the surface, be interpreted to be more favorable than that of the national office market. While the 1980s office market was characterized by speculative overbuilding, increased global competition was forcing a restructuring of industry that would discourage speculative building activity. Furthermore, the nature of industrial plant construction is more toward owner-specific and pre-leased facilities.

Out of the nine Census divisions, the building and expansion of industrial plants in the last three years has predominantly favored the South Atlantic, East North Central, and the West South Central states (Table 9–2). However, when viewing states individually, a rustbelt state, Ohio, was the leader in 1993 for receiving new and expanded facilities. The remaining four out of the top five were all sunbelt states.

As Table 9–3 indicates, industrial vacancy rates were markedly lower than those for office space. The total industrial inventory in suburban areas is more than twice that in central business districts, and this differential is expected to increase over time with continued suburbanization of industry. The movement towards flexible technologies such as just-in-time production systems with accompanying trends towards adjoining warehousing and distribution systems to minimize inventory costs, requires different facility designs. This suggests existing industrial properties may not be suitable. This may be even more the case with older industrial properties found in central cities. Given historical pre-auto building density patterns, these properties are more likely to be bounded on all sides and unable to expand.

Data on the Atlanta Metropolitan region ranked the nation's hottest industrial property market for 1994 appear to bear this out (*Dorey's Atlanta Industrial Guide*, 1994, p. 14). The overall vacancy rate in the Atlanta industrial market was 7.9% while

Table 9–2: New and Expanded Manufacturing Facilities, 1991–1993.

	New	*Expansion*
Top Three Census Divisions		
South Atlantic	1,111	1,170
East North Central	763	1,071
West South Central	599	758
Top Five States		
Ohio	355	594
North Carolina	375	467
Texas	356	319
Florida	237	215
Kentucky	133	332
U.S. Totals	3,880	4,495

Note: Facilities counted are those that either have: $1 million in investment, create 50 or more jobs, or result in at least 20,000 sq. ft. of new floor area.

Source: *Site Selection,* February 1994.

Table 9–3: The State of CBD and Suburban Industrial Markets, 1993.

Location	*Inventory (Sq. Ft.)*	*Vacancy Rate (%)*	*Vacancy (Sq.Ft.)*	*Avg. Ann. Net Absorp. 1990–1993[1] (Sq.Ft.)*	*Total Value Change 1990–1993[2] (%)*
Central Business District	1,643,566,469	13.70	224,622,288	22,587,066	–11.70
Suburban	3,649,700,792	12.00	437,683,472	38,945,743	–10.20

Notes:
[1]Average of annual totals of 41 U.S. metropolitan market areas.
[2]Sum of annual averages of 41 U.S. metropolitan market areas.

Source: Derived from Nunnink, K.K. (Ed.). (1994). *Viewpoint 1994*, Minneapolis: Valua-tion International, Ltd., February.

that for the central city submarket was 12.1% (*Metro Atlanta Industrial Broker Guide*, 1994).

Another way to view the production approach's influence on building design and layout is by its distinction as a "manufacture-to-order" versus "manufacture-to-stock" approach. The manufacture-to-order process is integral to the just-in-time plantwide production philosophy whereby production runs are set up from actual sale orders. The process entails low inventories because the product is only produced when an order has been received for it, and it is shipped to its customer upon completion. Thus the system requires relatively little space for inventory storage compared to manufacture-to-stock systems where a product is stockpiled until a future order arrives for it (Myers, 1994).

The difference in the two approaches can translate into extremely different building requirements as the following example illustrates:

> . . . the Japanese automobile industry applied [the manufacture-to-order] concept to a proposed facility layout that was essentially the same as a Ford Motor plant located in Batavia, Illinois. Both companies manufactured small- to medium-sized transmission gears, with their production output being virtually the same. The Ford plant contained 39 machines while the Japanese plant housed 41 machines, which were all essentially the same type of machinery. The only difference between the two plants was the Japanese were implementing the manufacture-to-order philosophy for production while the U.S. plant was set up for manufacture-to-stock. The Ford Motor plant in Illinois required 900,000 square feet for this facility, while the Japanese plant required only 300,000 square feet (Myers, 1994, p. 301).

This example might suggest JIT or manufacture-to-order approaches could be installed more easily in existing industrial facilities of central cities that tend to have a smaller footprint and/or have multiple stories than that which characterizes competing suburban space. This point can be illustrated with a brief account of an extreme case of just-in-time adoption that shows some promise for the use of existing, otherwise obsolete manufacturing structures. This case of just-in-time production was adopted by another Milwaukee-based firm, the Allen Bradley Corporation. Belonging to the same group of founding firms that included Allis-Chalmers and Schlitz, Allen Bradley was started in 1903 to make motor controls for industrial cranes. It was acquired by Rockwell International in the late 1980s. In 1992, Allen Bradley introduced a 45,000-square-foot "factory" within its existing multistoried facility on Milwaukee's South Side. This factory-within-a-factory builds circuit boards for industrial automation controls and systems. Its production cycle begins at 5 a.m. each day with orders conveyed by the company's mainframe computer that it received the evening before from sales representatives located around the world. It has a stockless production system with next-day shipping, and therefore no real need for warehousing space. Further, its location meets the public infrastructure requirements of today's industries discussed below.

The public infrastructure requirements ranked highest today by type of facility do not necessarily rule out central cities as viable location contenders. A 1993 survey found access to interstate highway among the five key factors for manufacturing plants and distribution/warehouse facilities, while access to a major airport was among the top five for regional headquarters and sales offices (Coffee, 1994). These factors at least do not work against central city location choices, and could work in their favor.

For example, a marketing guide for the Atlanta central city area suggests that current inventory provides a wide variety of facilities for "cost-conscious industrial concerns seeking a central location in close proximity to Atlanta's extensive interstate highway network" *(Dorey's Atlanta Industrial Guide*, 1994, p. 67). If overall manufacturing space requirements are shrinking through the adoption of just-in-time production processes, then the low-cost vacant central city plants could be competitive with suburban sites, much like the rationale that drove the Schlitz office park developers. And clearly such sites could be at least, if not more, competitive than more isolated rural sites. Finally, as the Schlitz and Allis-Chalmers examples illustrate, this stock of abandoned industrial plant can be put to adaptive reuse. The potential is great, in particular, for warehouse conversion. The wood floors, high ceilings, and brick interiors frequently found in inner city warehouses have appeal for the residential, office, and specialty use markets (i.e., art and music studios). However, the next section ,which discusses redevelopment challenges, will make clear that a far more explicit policy will be required to counteract suburban shifts in demand for office and industrial structures.

REDEVELOPMENT CHALLENGES

Preference for rural or suburban greenfield industrial sites has been a primary location factor throughout much of the postwar period. The widescale adoption of assembly-line production systems shifted the demand to large, single-story manufacturing plants. That the majority of the labor force shifted to driving private autos to their manufacturing employment sites also required large amounts of space for parking lots. These were the primary factors contributing to the widescale abandonment of industrial infrastructure in central cities. Increasingly, we have seen the abandonment of office infrastructure as well. Following the rapid postwar suburbanization of housing, office sector employment has shifted to suburban locations that are closer to the labor pool. Advances in information technologies such as electronic mail and data transfer and fax machines have facilitated the "decoupling" of back office functions such as accounting and insurance claims processing from executive office functions allowing corporations to take advantage of less expensive building sites and the reserve labor pool of suburban women.

The chronicles of Schlitz and Allis-Chalmers discussed earlier suggested that localities need not give up on abandoned industrial sites, for even within the core of our old and seemingly obsolete industrial cities, the evolving mode of production has use for industrial wastelands that can result in economic rebirths where Fordist extinctions occurred. However, both of these redevelopment efforts were initiated in the early to mid-1980s. Since then, an important qualifier must be added to this optimistic assessment. The increased pressure for environmental remediation of industrial contamination serves to impede the replication of these success stories for other locations. In fact, the developer of Schlitz Park has stated that in today's climate, he would not have been able to go forward with plans to create the office park.

One of the legacies of the industrial economy's impetus toward flexibility has been to leave behind a substantial barrier to economic development for those areas in the United States in greatest need of revitalization. The term "brownfields" has been given to these areas to distinguish them from the greenfield designation of undeveloped urban fringe or rural land. The barrier to brownfield redevelopment is the legacy of environmental contamination on sites and their surrounding areas which industry has subsequently abandoned. In particular, the barriers to development that environmental contamination possess for the central cities of our oldest industrial areas may relegate substantial portions of the urban core permanent wasteland if present environmental policies and economic development practices continue. With their continuation, the prospects for job creation and economic development prospects for the population left behind in America's central cities may dry up altogether (Leigh, 1994).

The magnitude of environmental contamination across the nation can not be fully calculated but appears to be extensive. In addition to the 1,250 or so sites on the National Priority List of the Environmental Protection Agency (EPA), Salvesen (1993) cites estimates of 30–40 thousand contaminated industrial and commercial sites throughout the U.S. Heavy industrial facilities such as steel mills, as well as chemical and other types of processing plants, are well-recognized sources of environmental contamination, but ubiquitous service-oriented facilities such as gas stations, auto repair shops, and dry cleaners have also been prime contributors to the pollution of land, water, and air (Bartsch et al., 1991). The EPA has estimated that a quarter of the nation's several million underground storage tanks holding petroleum or other hazardous substances could be leaking and contaminating the surrounding soil and ground water. Additionally, the historic practices of offsite disposal of industrial waste in land-fill sites not designed to contain hazardous and toxic chemicals contributes to the pervasiveness of contamination problems.

As has been observed elsewhere (Leigh, 1993), the present level of awareness of the extent of environmental contamination in brownfields can be seen as a double-edged sword: what is not known could be dangerous, but what is could prove harmful as well. For example, in not knowing that a brownfield site has environmental contamination the population may unwittingly be exposed to serious and undetected health hazards. However, actual knowledge of a brownfield site's contamination more often than not results in that site no longer being considered viable for redevelopment, and can even taint the prospects for contiguous sites. Under our present environmental regulations, this is due to the liability that owners of a contaminated site assume by law whether or not they were the actual contaminators of the site, and to the costs which may not be fully calculable at the outset that would have to be incurred to clean up the site for reuse.

We have "environmental redlining." Developers and banks will not consider properties with previous uses associated with contamination. Risk analysts, compiling statistics on the frequency of contamination of sites based on former uses, suggest that sites with previous industrial use will have a 65–99% reported frequency of contamination (Noonan and Vidich, 1992). This "environmental redlining" could significantly dim the economic development prospects for the population residing in the vicinity of the site. Those most negatively impacted are the same black and other minority population groups found in the inner city whose discriminatory experiences with the residential real estate and mortgage industry led to the original coining of the term.

Even when a contaminated site has been "cleaned" and the contamination barriers to redevelopment have been eliminated, it appears that aversion to the site cannot always be eliminated. For example, Johnson Controls Incorporated, a Milwaukee-based multinational firm whose primary products are automotive batteries, building environmental controls, automotive seating, and plastic bottles, has been unable to find a buyer for the Atlanta site of one of its former lead acid battery plants despite the fact the company has cleared and cleaned the site completely and the site is located in one of the hottest markets in the nation (*Dorey's Atlanta Industrial Guide,* 1994). The site should be a prime commercial property as it is located on a corner lot of a major thoroughfare, has essentially been brought back to "greenfield" condition, its building structures have been razed, and the State of Georgia's requirements for soil and groundwater remediation have been satisfied. However, buyers are still wary of the site and are unwilling to acquire the cleaned site unless the corporation will warranty the cleanup work that has been done (despite the State's signed-off on the work). Buyers even want an agreement that Johnson Controls would be responsible if future environmental contamination problems were found with the site. From Johnson Controls' perspective, this is unacceptable since it has no control over the actions of the buyer of its property: the possibility exists that the buyer could add contaminants to the site similar to those that were added during Johnson Control's period of ownership for which Johnson Controls would inaccurately be assessed blame and cleanup responsibility (Beaudoin, 1994).

Private landowners such as Johnson Controls are trapped by evidence of, or the perception of, environmental contamination. However, increasingly, the public sector is caught in this "entrapment." Tax delinquency of suspected properties is a growing problem for municipalities who, in turn, are unwilling to foreclose on such properties for fear of becoming liable for clean-up. The reluctance to use the foreclosure weapon acts as a *de facto* license to holders of suspect properties to stop paying their property taxes. For example, the number of tax delinquent commercial or industrial properties on which the city government of Milwaukee, Wisconsin chose not to foreclose because of the possibility of contamination grew from 175 to 474 between 1992 and 1994.[3]

The overwhelming impetus has long been toward the development of greenfield sites. This has resulted in a continued shrinkage of the nation's primarily agricultural and virgin lands, continual demand for new office and industrial infrastructure, and corollary decline in existing urban infrastructure. The environmental constraints to brownfield redevelopment presently serve to strengthen an impetus that does not need reinforcement.

FIXED STRUCTURES IN TRANSITION: SUMMARY AND PROGNOSIS

This chapter has attempted to focus on an array of trends that are affecting the use of fixed structures, specifically, office and industrial facilities, in what is perceived as a heightened era of flexibility in the capitalist economy. The overall trends do not bode very well for the fate of central city economies. There is very little that suggests the tide of abandonment of older office and industrial buildings will be turned. Industrial capitalism has long relied on its ability to freely shift locations, abandoning redundant sites for new sites, consuming increasing amounts of greenfield land in the process. In the late twentieth century, we are increasingly forced to contend with this downside of capitalist production.

This writer tends to agree with those who contend that the search for flexibility in capitalist production is not a new phenomenon. According to Sayer and Walker:

> What is therefore needed is not an obsessive focus on flexibility but a broader wareness of new forms of the division of labor and new ways of organizing those divisions (Sayer and Walker, 1992, p. 199).

Just as necessary is an equal awareness of how the new forms and ways of organizing the production process and labor inputs affect the demand for office and industrial infrastructure.

Office and industrial buildings cannot get up and move with shifts in the location of production, and the negative externalities associated with abandoned buildings are great. That is, the surrounding neighborhood suffers from increased vandalism, deteriorating buildings become eyesores, they yield lower property taxes and lower the value of neighboring real estate. The lack of economic use of such buildings decreases employment opportunities for the area's population, particularly if that population is low-income and has limited mobility. Environmental contamination spills over to affect property values and desirability of adjacent properties.

Further, the shift of, first, industrial and, now increasingly, office locations away from a metropolitan area's central business district has generated a multicentric pattern of (sub)urbanization. Accompanying this multicentric form are inefficient commuting patterns that largely require the use of private autos and significantly contribute to air quality problems. Also, this multicentric pattern of (sub)urbanization creates increased pressure to devote limited public resources to new suburban infrastructure improvements and decreases that available for urban infrastructure redevelopment.

Unintentionally, state and federal regulations intended to prevent or remediate environmental contamination have resulted in large-scale market failure for brownfield redevelopment projects. In meeting their professional responsibilities while avoiding liability exposures, environmental scientists along with financial and real estate analysts have created a climate of environmental redlining for a substantial portion of the nation's urban core, dimming even further the prospects for its resident underclass.

In our present regulatory and real estate financial climate, there is little or no incentive for reusing existing infrastructure. In fact, the disincentives are substantial, and existing infrastructure is a disposable commodity. In contrast, Adams and Parham (1995) have noted that the European building community does not consider its products disposable commodities.

The search for flexibility in capitalist production is not new. What would be new, and a major advance in central city revitalization, is the American development community's internalization of flexibility that has been attributed to European practice:

> The United States must also build for the long term, incorporating flexibility for change *internally* rather than continually replacing its cities and architecture (Adams and Parham, 1995, p. 43).

ENDNOTES

[1]This phrase appears in Storper (1989).

[2]Data provided by Schlitz Brewery Annual Reports, 1962 to 1981.

[3]Data for 1994 obtained from the City of Milwaukee's Tax Assessor.

REFERENCES

Adams, H.L. and Parham, D.W. (1995). "Repositioning the Office Building." In *Reinventing Real Estate*, Washington, DC: Urban Land Institute, pp. 36–43.

Bartsch, C. et al. (1991). *New Life for Old Buildings: Confronting Environmental and Economic Issues to Industrial Reuse*. Washington, D.C.: Northeast-Midwest Institute.

Beaudoin, J., Manager of Environmental Relations, Johnson Controls Battery Group, Inc., interview with author, 21 April 1994.

Bell, M. (1993). "The Virtual Office: Hit or Myth?", *Site Selection*, Vol. 38, No. 6, pp. 1296–302.

Coffee, H.E. (1994). "Location Factors: Business as Usual, More or Less," *Site Selection*, Vol. 39, No. 1, pp. 34–8.

Dorey's Atlanta Industrial Guide, 3rd Quarter, 1994, Atlanta: Dorey Publishing and Information Services.

Grunau, G. President of Grucon Corporation, interview with author, Milwaukee, WI, 15 November 1991.

Harvey, D. (1989). *The Conditions of Post Modernity*. New York: Basil Blackwell.

Kass, M. (1992). "Grand Avenue TIF to Cash Out Early," *The Business Journal*, Vol. 9, No. 21, pp. 1-2.

Leigh, N.G. (1993) "The Making and Unmaking of Industrial Wasteland: Economic Development Insights from Two Milwaukee Chronicles," *Economic Development Commentary*, Vol. 17, No. 1, pp. 4–11.

_____ . (1994). "Introduction to Environmental Constraints to Brownfield Development," *Economic Development Quarterly*, Vol. 8, No. 4, pp. 325–8.

Metro Atlanta Industrial Broker Guide, Spring 1994, Gaithersburg, MD: Black's Guide, Inc.

Myers, J. (1994) ."Fundamentals of Production that Influence Industrial Facility Designs," *The Appraisal Journal*, Vol. LXII, No. 2, pp. 296–302.

Nelson, K. (1986). "Labor Demand, Labor Supply, and the Suburbanization of Low-Wage Office Work." In Scott, A.J. and Storper, M. (Eds.), *Production, Work, Territory,* Boston: Allen & Unwin, pp. 149–71.

Noonan, F. and Vidich, C. (1992). "Decision Analysis for Utilizing Hazardous Waste Site Assessments in Real Estate Acquisitions," *Risk Analysis,* Vol. 12, No. 2, pp. 245–51.

Nunnink, K.K. (Ed.). (1994). *Viewpoint 1994.* Minneapolis: Valuation International, Ltd., February.

Ogilvie, H. (1994). "This Old Office,"*Journal of Business Strategy,* Vol. 15, No. 5, pp. 26–34.

Reich, R.B. (1992). *The Work of Nations.* New York: Vintage Books.

Salvesen, D. (1993). "Contamination and the Reuse of Land," *Urban Land,* December, pp. 32–6.

Sayer, A. and Walker, R. (1992). *The New Social Economy: Reworking the Division of Labor.* Cambridge, MA: Blackwell.

Schlitz Brewery Annual Reports, 1962 to 1981.

Storper, M. (1989). "The Transition to Flexible Specialization in the U.S. Film Industry: External Economies, the Division of Labour, and the Crossing of Industrial Divides," *Cambridge Journal of Economics,* 13, pp. 273–5.

Wendel, C.H. (1988). *The Allis-Chalmers Story.* Sarasota, FL: Crestline Publishing Co.

10

PERIOD AND PLACE, CAPITALIST DEVELOPMENT, AND THE FLEXIBLE SPECIALIZATION DEBATE

Kevin R. Cox
Department of Geography, The Ohio State University, Columbus, OH 43210-1361 USA

INTRODUCTION

This chapter concerns the way in which we use concepts of period and place in social theory. It is prompted by thoughts raised in the context of the debate about flexible accumulation (FA) and flexible specialization (FS). The idea of industrial divides—whether that first promulgated by Piore and Sabel (1984) between mass production and a more recent revamped craft production or the military-industrial divide of Markusen (1991)—serve to emphasize the issue of periodization in social theory. The attendant concern with industrial districts of the sort first identified by Piore and Sabel brings to mind the issue of just how and why we might divide up substantive domains geographically. Some of the critique of flexible accumulation and flexible specialization theses has also served to draw attention to differentiation in the geographic plane: concerns for the institutional distinctiveness of Japan come to mind here (Kenney and Florida, 1988; Sayer, 1989; Curry, 1993).

Questions of period and place are closely related, but not identical, to those of time and space. Curiously, the FA/FS literature has little to say on these topics. Alongside the literature on new industrial spaces and production complexes, for example, there is another that focuses on so-called globalization, the effects of the New International Division of Labor on more developed countries as in the creation of numerous rust belts (Gertler, 1992), and the concomitant growth of world cities: a literature, in other words, that instead of drawing attention to the fragmented, discontinuous nature of space implicit in FA/FS theses is concerned with its integrating character. A similar lacuna applies to the question of time. Just as capitalism has a spatially integrating/expansionary moment, so too is it expansionary and integrating in the temporal domain: exhibiting strong tendencies toward cumulative development and a concurrent colonization of the future. What difference does recognition of these practices make to an evaluation of the FA/FS debate?

Period and Place

In order to grasp the differentiations of the capitalist world a common approach has been to divide it up along both historical and geographical dimensions. It is common to refer to a problem of periodization. Regulation Theory seeks to identify periods distinct with respect to the institutions through which accumulation has been secured. Piore and Sabel distinguish periods according to their dominant technological ensembles, their proposed meta-history consisting of an alternation of mass production and craft production paradigms.

The contemporary interest in new industrial spaces or territorial production complexes serves to remind us of attempts to classify in the geographical domain. There are scale elements here as there are in world systems discussions of periodization. Indeed in his discussion of post-Fordism Sayer (1989) has suggested that there is more coherence in institutional ensembles within countries than there is within periods:

> . . . the superior industrial performance of Japan or the FRG over the USA and UK is not so much a question of their relative commitments to mass production or flexible specialization but to do with broader, "environmental" characteristics which cut across labour process and production scale distinctions. In this respect comparative studies may serve us better than binary histories (p. 691).

These distinctions and the projects which inform them echo more general calls for context-specific social theory. Given that as the goal, however, the problem is: What criteria should one make explicit in the attempt to derive concepts of period and/or place? And in that light, are (e.g.) the periods identified by Regulation Theory or by Piore and Sabel the "correct" ones? This author's criteria are realist ones.

In the first place we need categorizations that can shed light on the necessary properties of agents: of firms, workers, and state agencies. This implies a preference for what Sayer (1992) has described as "rational abstractions": concepts, that is, which bring together the internally related and discard the incidental. Concepts like Fordism, competitive regulation, flexible accumulation, and flexible specialization need to be critically viewed from this standpoint if we are to hope to use them for analytic purposes. In particular there has been a lot of scepticism regarding the concept of flexible specialization. Critics have drawn attention to the non-necessary connections between such defining moments as: firm size; the degree to which equipment is dedicated to particular tasks; and length of production run. The same goes for FA where the relations within specific firms between functional and numerical flexibility seem thoroughly contingent (Sayer, 1989; Williams et al., 1987; Pollert, 1988). The concept of Fordism has not been subjected to the same sorts of criticisms. But the looseness of the correspondence between the Keynesian-welfare state on the one hand and the prevalence of fordist labor processes on the other does not suggest it is any less "chaotic" a concept.

The second criterion of the defensibility of concepts of place and period is the degree to which they acknowledge the fundamental presence and power of the social relations of a capitalist society and what those relations imply for essential social powers and needs. We are, after all, talking about periodizations and place-partitions of capitalism. Whatever the relations specific to particular periods or places they will be capitalist relations and so subjected to the constraints and opportunities peculiar to that mode of production. They are, rather, more concrete instances of those relations and we should not lose sight of that fact. More generally we should be thinking in terms of hierarchies of abstractions (Cox and Mair, 1989) though it needs to be emphasized that the more concrete does not necessarily correspond to the place- or period-specific. Some corporations are arguably global in their presence and endure from one period to another; but they still provide more concrete instances of the capitalist corporation. Different branches of production and their various subdivisions also crosscut the specificities of time and place imposing (e.g.) their own technological requirements. All this would suggest that defensible concepts of place and period require a lot of hard conceptual work.

Time and Space

Social relations, including those of capitalism, occur not only within periods and places but also between them. Giddens's (highly abstract) concept of time-space distanciation captures some of this. Contemporary discourse about Third Italies, new industrial spaces or industrial districts, for example, coexists, rather uncomfortably in some instances, with another. This focuses on the creation of New International Divisions of Labor, the concomitant formation of world cities, and a more all-embracing concept of globalization. A common explanation of the demise of Fordism looks to processes of this nature. Lipietz's idea of peripheral Fordism (1987, pp. 78–81) and the externalization of the production-consumption relation by NICs in a context where regulatory institutions are lacking at the global level is indicative. That is not to imply, however, that there is anything "new" about globalization in the way in which some current ideas might suggest. The world market is as old as capitalism though the concrete forms through which that integration has been expressed have changed: this much is apparent in the term "New International Division of Labor" and its implicit contrast with an older and different one.

Likewise, capitalism has always colonized the future. The obvious way in which this has occurred has been through its technological and organizational dynamism and subsequent claims to the product of (future) social labor. This in turn has resulted in attempts to control a (necessarily) changing environment so that those claims can be realized. As with the stretching of relations over space so there have been concrete changes in the way the future is integrated into the present. This is underlined by the tendency for innovation, re-making the firm, and planning for the future to become increasingly institutionalized. In the course of their development, moreover, firms discover new forms of organization which can buffer them against the effects of the unanticipated, as in the joint stock company or the multilocational firm: new forms which may become more widely adopted and, in some cases, necessary conditions for business strategies for a very long time into the future. This suggests that just as we should view with scepticism attempts to partition capitalist development geographically so we should be similarly suspicious of the discrete periodizations of the Regulationists.

THE NATURE OF CAPITALIST DEVELOPMENT IN TIME AND SPACE

Critiquing Concepts of Period and Place

There is indeed substantial critical leverage to be gained by placing periodizations and place-partitions in the context of the long-term developmental tendencies, spatial and historical, of capitalism. Brenner and Glick (1991) have recently provided an exemplary critique of the Regulation theorists along these lines and provide a good starting point. The fact that their critical rebuttal is accompanied by quite devastating statistical materials significantly bolsters their case.

Central to the Regulationists' conceptual armory is the idea of a mode of regulation. This consists of a set of institutional structures (state legislation, norms, modes of organization) which govern interfirm and business-labor relations. These institutional structures function to facilitate a correspondence between the respective transformations of production and consumption.

Brenner and Glick draw a major contrast between a regulatory structure dominant in the nineteenth century and which, according to Regulationist arguments, gave capitalist accumulation a mainly extensive character; and a regulatory structure that was progressively implanted after the Second World War, imposing on it an enduringly intensive character. An intermediate phase is characterized by intensive accumulation on the basis of scientific management and the assembly line; but also by regulatory institutions incapable of overcoming chronic tendencies toward underconsumption. The regulatory structures characteristic of the nineteenth century are termed "competitive." It is their competitive nature that is believed to have prevented capitalist accumulation from moving beyond an extensive character based on the extraction of surplus value in its absolute form and a movement to new geographical frontiers to an intensive one in which the extraction of relative surplus value became dominant. Those structures of the more recent Fordist period which allowed the suspension of the (supposed) problems of underconsumption of the 'twenties and 'thirties are characterized as "monopoly."

One of the key points of the Regulationists, therefore, is that capitals in the nineteenth century were supposedly unable to move beyond a regime of extensive accumulation. They had difficulty, therefore, in arriving at a virtuous circle of technological development, the cheapening of commodities, and a raising of material standards. This is laid at the door of their dominantly competitive regulatory institutions. Competitive labor markets limited purchasing power and created disincentives to invest in productivity-raising methods. Intense competition between firms added to these disincentives by increasing the risk attached to investment in fixed capital.

Brenner and Glick counterpose to this the implications of the underlying, production-relations driven logic of the capitalist mode of production. A competitive environment may certainly make investment in new methods/techniques risky but the refusal to invest is perhaps equally risky. Moreover, the risk is to some degree offset by the possibility of superprofits for the innovators. And not all productivity-raising methods are capital-intensive but involve reorganization of the division of labor within and between firms and exploiting the usefulness of cooperation within the factory. At the same time, the shift to relative surplus value, by cheapening goods, should allow an increase in real wages and an extension of the market, assuming, that is, that nominal wages and employment remain steady:

> It has, therefore, seemed an elementary empirical generalization that, for the history of capitalism the rise of mass production does not require but rather issues in, mass consumption—that the latter has depended upon the former, even though, in some important ways facilitating it (p. 66).

There is no need, therefore, to posit institutional discontinuities in order to understand the developmental trajectory of capitalism. Furthermore, just as technology in the hard sense developed in an ongoing way by virtue of capitalism's property relations, so too did technology in the soft sense: the sorts of institutions that regulate risk, facilitate investment and consumption. The growing needs for fixed investment led to a plethora of innovations on the institutional front including: the joint stock company, regulated banks, and vertical and horizontal integration. And from the consumer's viewpoint the emergence of such institutions as building societies or savings and loans, the development of the mortgage loan and of the insurance industry are of signal importance. The introduction of these organizational innovations (like new technologies in the narrow sense) has been continuous and not discontinuous in the sense implied by Regulation Theory.

There are other secular tendencies to be considered. The Regulationist counterposing of an earlier competitive phase to a later phase of monopolistic regulation is especially puzzling. The argument is that the displacement of small, competing firms by larger, more market-dominant firms facilitated control over prices and hence over markets while at the same time reducing the risk of heavy fixed investments. Yet there are strong arguments suggesting that far from capital becoming more monopolistic and so displacing an earlier era of high competition it may be, in fact, that it is the reverse. Brenner and Glick echo Harvey (1982, Chapter 5) when they emphasize the gradual elimination of locally protected markets as a result of the development of the means of transportation; and the development of banking institutions which has allowed firms to invade new markets and branches of production in a way which would have been difficult in the nineteenth century. Weeks (1981, Chapter 6) comes to the same conclusions. The rise of the NICs and their invasion of the markets of the developed countries is only the latest phase in a long history: a history, that is, which is inscribed in the logic of the capitalist mode of production. As Brenner and Glick stress,"[T]he general weakness of Regulation Theory . . . is its failure to take adequately into account the broader system of capitalist social-property relations that forms the backdrop to their succession of institutionally defined phases" (p. 105).

What, however, of the capitalist space economy and its partitioning into places: nation states, territorial production complexes? Is it possible to develop a similar critique? Certainly it is not difficult to find counterposings of the top-down, systemic logic of the space economy to the bottom-up arguments of the various advocates of analyses which are more place-sensitive. Lipietz (1993) has recently provided something of an inventory of these. On the one hand there are those approaches to understanding place-space relations which focus on the internal structure of regions. And on the other are those which define regions by their place in a more total or encompassing structure. In this way he contrasts stage theories of economic development which abstract from more global relations, the varied contributions of Rostow, Colin Clark, and Raymond Vernon, with the critique of dependency theory and the development of underdevelopment. More recently still there is the antinomy of the New International Division of Labor approach to the changing space economy and that of the advocates of new industrial districts, as in the case of the Third Italy or Silicon Valley, and national (i.e., place-specific) regulatory institutions. But though the con-trasts are drawn there is little in the way of resolution. In fact references to "an impassable dialectic" of the global and the local suggest that one has to be pessimistic about its possibility.

This author thinks that one can move beyond this and develop a more adequate sense of "the difference that place makes." This author also thinks that while Brenner and Glick press home some telling points there is a sense in which they do not do justice to the more concrete aspects of capitalist development in their temporal frame, though in fairness the concern here lies beyond their particular brief of a critique of Regulation Theory.

There are a number of very important issues here having to do with what we signify when we use concepts like time and space and period and place, and all of them in the context of a discussion of specifically capitalist development processes. More than a simple contrast of continuity vs. discontinuity is involved. There are, for instance, questions of scale. "Place" signifies something relatively local in scale and so is sometimes contrasted to "global" as a larger scale. This seems at least implicit in Lipietz's (1993) discussion. Similarly "period" may be contrasted with longer-term processes like "capitalist development," as seems to be happening to some degree in the Brenner-Glickman paper.

To add to the confusion there is often an unfortunate conflation of different scales with different levels of abstraction. The global may be assumed to be the domain of more abstract social relations while more concrete structures are reserved for the local level, as indicated by the idea of the specificity of "place." Yet global structures may be of a very concrete character; multinational corporations are different from one another. Likewise, local structures of social relations can be very abstract and so replicable from one "place" to another, as in local government structures. Likewise "periods" are far from different in all their essentials; one can abstract from their concreteness to identify social relations which they have in common. But by the same token the longer term may acquire its own specificities as new organizational innovations are adopted and provide an essential condition for all that follows. This is the notion of what the economic historians have called the path dependence of economic development, though as we will see this is also substantially tied up with the existence and significance to capitalist development of another kind of dependence: place dependence.

Place-dependence and Path-dependence

The continuity of capitalist development over space and time provokes images of flow, of effortless reproduction, and of (if not endless, then certainly expanding) frontiers of substitutability. These images, of course, are quite onesided. Capital can only move from one point to another, whether in space or in time, if it is actually appropriated; and in order for it to be appropriated it has to be produced. This occurs at a particular point in time and space and as a result of a coordination of diverse activities at equally diverse points in time and space: a coordination of activities in the sense implied by Hagerstrand's time-space geographies.

Production, and the linkages which are the conditions for it, have a fixity. This fixity varies. So called fixed capital may be moveable, as in the case of some machines. But infrastructures like steel mills, chemical plants, and factories, not to mention the housing of the workers, and the roads along which the various commodities entering into and leaving the production process move are physically embedded in the landscape.

Activities have to be coordinated and not just in the sense of recruiting and bringing workers of the appropriate skills together in one location. Supplies of raw materials or intermediate products have to be arranged. In some instances these may have to be mediated, perhaps by face-to-face interaction and mutual understandings. Interaction with customers, perhaps on the design of products, may be necessary. In addition to physical infrastructures for production, therefore, it may be helpful to talk also of the social infrastructures through which production is mediated: local knowledges, local codes of behavior, local labor markets, external and internal (Storper, 1994).

This means that production may become embedded in particular places and dependent upon the physical and social infrastructures characteristic of those particular places. The physical infrastructures are absolutely immobile and represent value of long life whose abandonment is costly. Social infrastructures are in principle mobile but it is rarely easy to coordinate all the various moves if the infrastructure is to be re-established somewhere else. Much depends on the degree of standardization of the infrastructures in question. Some are relatively modular in character as in the case of so-called branch plant towns. But others are clearly unique and include not just the Silicon Valleys of the world but also one-industry towns where production, while more vertically integrated, is still dependent upon a core of labor skills difficult to replicate elsewhere.

On the other hand, and as per remarks made at the end of the last section, it is not desirable to reduce "place" and "place-dependence" to the idea of the "local." One way of approaching this is through a recognition of scale elements in the concept of "place." Discussions of so-called short-termism in the context of problems of economic growth in the U.S. and in the U.K. have served to give focus to some of the institutional differences between countries. This is something that Sayer (1989) has referred to in his critical evaluation of post-Fordism in the light of the continuing success of mass production in Japan. One of the conditions for this appears to be modes of organizing inter-firm relations, i.e., social infrastructural arrangements, which are quite distinctive to that country. These can serve to embed quite as much as the more local elements of place-dependence.

However, still more is involved than this. Agents may become embedded not just in one place but in a network of places. These networks again can occur at a variety of different geographic scales. The new spatial divisions of labor of firms can assign particular roles to particular plants in different localities, each plant providing an essential component for the final product and being difficult to substitute for, particularly over the short term. One thinks here of the relations between the Japanese auto companies and their subcontractors: relations built up over a long period of time and attached to which, therefore, are substantial opportunity costs. Similarly, firms invade new markets but that does not mean that they can leave them at will and penetrate new ones. Successful marketing in new localities, countries, or whatever, involves the development of a marketing infrastructure: outlets, service and repair facilities, a brand image, all of which imply sunk costs of a considerable magnitude. Furthermore, to the extent that firms are successful in their respective markets, this serves to further entrap since the barriers to entering new markets thereby become more formidable.

Just as there are place dependences, however, so there are some relations that are more easily substitutable in a locational sense. How significant they are will obviously vary. It will vary, among other things, with the degree to which inputs, including labor inputs, are standardized. It will also vary with the standardization of linkages with other firms and with the specialization/modularity of the physical facilities (e.g., steel mills vs. distribution centers or back offices). It is standardization which has allowed, among other things, for the creation of new spatial divisions of labor and the relocation of low-skill production jobs to the periphery which are one feature of contemporary globalization. Yet, and again, this locational substitutability is not something that can be tied down to any particular geographical scale. Firms may enjoy a good deal of substitutability between locations within a particular country but not between. This means that the global in an absolute, as opposed to relative sense, needs to be approached with caution. Even at the scale that is conventionally regarded as global, substitutability may exist for a firm but be a qualified substitutability: of New York for London for Tokyo, but not of New York for St. Louis.

Similar concerns structure this author's consideration of path dependence. This is an idea that has emerged from the work of evolutionary economists in attempts to understand the development of technologies. A central focus of their concern has been the way in which particular technologies assume a dominance over others even though there might have seemed little objectively at the start to recommend one over another. It would seem, however, to have applicability beyond the narrowly technological to questions of social organization and the meshing of concrete structures of social relations with the technical.

The idea is quite simple. It can be illustrated with the case of two competing technologies, both of which yield returns to scale. One, however, gains an advantage resulting in a monopoly domination. This allows it to dictate the path of technological development subsequently taken, even though it may turn out in hindsight to have been a blind alley. The advantage initially gained may be due to some random event: a technological breakthrough, perhaps, fortuitous government support, a particularly favorable environment of externalities, but once the advantage is achieved positive feedbacks take over and assure its relative growth and future dominance.

These positive feedbacks can assume a number of different forms. Customers learn how to use a particular product and so are likely to replace it with one of the same kind. Networks of repair facilities may come into existence along with supporting physical and social infrastructures so that a further bias is accorded the (emergently) dominant technology. The result is not inevitable and has no inherent efficiency properties. But once it has come into being it tends to channel further development along the particular pathway it has defined: particular knowledges, technical and financial, particular materials and skills in working with those materials, and particular slants in training.

Douglass North (1990) has tried to appropriate this idea for the understanding of institutional development. Further interest lies in the fact that the specific question he is trying to answer has substantial geographic content:

> What accounts for the survival of societies and economies that are characterized by persistent poor performance? . . . The implications of the (Darwinian) theory are that over time inefficient institutions are weeded out, efficient ones survive and thus there is a gradual evolution of more efficient forms of economic, political, and social organization . . .why would the relatively inefficient economies persist? What prevents them from adopting the institutions of the more efficient economies (p. 92).

The appropriation does have a certain plausibility, even if in North's presentation it is innocent of important political effects. Pathways of institutional development are not benign: they are always for somebody and not necessarily for everybody. The sequence of events that culminated in apartheid can be analyzed precisely in terms of the logic of choice identified by the evolutionary economists and the gradual elimination, by a variety of positive feedbacks, of more democratic alternatives. Once a developmental path is set on a particular course, the network externalities, the learning processes of organizations, and the historically derived subjective modeling of the issues reinforce it. To the extent that that pathway of institutional development has undesirable economic effects, the participants will construct an ideology that not only rationalizes the society's structure but also accounts for its poor performance. The externalization of blame, as in many Third World situations, suggests the plausibility of this.

This by no means exhausts the topic, of course. The notion of path dependence can, it would seem, be fruitfully applied not just to technologies or to institutions but to ensembles of industries and consumption practices. In an early paper designed to understand the energy-intensive nature of the American city, Walker and Large (1975) emphasized ideas of mutually reinforcing investment decisions and expectations on the part of the auto and oil industries and developers and the home finance industry. The inertia of fixed investments lent further impetus to what they described as a process of "spiraling causality" and the entrenchment of an automobilized society of low-density suburbs.

Moreover, there is nothing in the concept of path dependent development that would restrict it to developments within the sorts of periods identified by the Regulation Theorists, or to any particular span of time for that matter. Firms develop strategies in terms of what is available at a particular point in time and, to the extent that the strategy is effective, reproduce those conditions or functional substitutes for them as a necessary aspect of their growth trajectory. These conditions can be part of the long-term institutional development of capitalism or ones with more recent temporalities. This embeddedness imposes constraints on the abilities of firms to integrate the novel. Accordingly, very different organizational-technological ensembles may coexist under the same broad umberella of a particular regulatory regime and indeed challenge its intellectual coherence.

Place- and path-dependence are transparently related. Particular places are characterized by particular types of path-dependence and agents are not just locked into a particular developmental path but also into that particular place where their (path-dependent) powers can be exercised. This follows from the sorts of externalities and social infrastructures characteristic of particular areas and their (specialized) physical infrastructures. But place- and path-dependent development are not invulnerable. Only if capital continues to flow through their social relations can they survive. The standardization of particular industrial processes driven by the attempt to cheapen products and broaden final markets can be a serious threat. As the literature on new spatial divisions of labor emphasizes, the de-skilling of labor processes may allow the dispersion of production to more peripheral locations resulting in a "hollowing out" of local or national economies and in the end to particular forms of path-dependent development.

Place- and path-dependent development is attractive to firms since it provides them with the possibility of super profits: profits derived from place-specific advantages which cannot be replicated elsewhere and rooted in a historically specific concatenation of technical, organizational and socio-political conditions. In this regard it is a form of what Storper and Walker (1989) have called "strong competition." They contrast this with "weak competition." In weak competition the wide variety of substitutable locations destroys the possibility of super-profits. What should be underlined here, however, is the relationship between these two forms of competition.

Strong competition has, as one of its necessary preconditions, weak competition. It is weak competition, through its levelling out of profits, that provides the incentive for capitals to develop and exploit new pathways to accumulation: new pathways which can only be replicated elsewhere and by others with great difficulty. On the other hand, strong competition tends to give way, though not inevitably, to weaker forms. This occurs as firms try to cheapen the product in an attempt to create mass markets for it.

Place- and Path-dependence in Context

One can now to situate place- and path-dependent forms of capitalist development with respect to a broader context. For every place- and path-dependent form has to satisfy the systemic requirements of the capitalist mode of production. It must be capable of facilitating the valorization of capital in a context of inter-firm competition and contradictory class relations. Not every one of these can pass this test. As Brenner and Glick (1991) argue:

> Paths of capital accumulation . . . are simply incomprehensible apart from a specification of the broader system of social-property relations in which they are embedded. This is not only because those systems will define the range of economic strategies that individual actors can find it sensible to adopt; it is also because those systems constitute a field of natural selection for the adoption of institutions and heavily determine the effect of given institutions, once adopted, on capital accumulation (p. 110).

Brenner and Glick also define this field of natural selection in explicitly geographic terms. In their view the fact of the world economy makes even more implausible the project of conceptualizing the history of capitalism as a progression of institutionally determined, nationally situated, modes of development. Rather:

> . . . the given international distribution of productive power will have a central role in determining what institutions are even viable within national economies at a given historical juncture as well as what will be their effect on capital accumulation since unless they are shielded in some way these institutions must directly respond to international competition (p. 111).

In these ways place- and path-dependent forms of development are selected in and out. The same applies to a wide range of institutions and technological-organizational ensembles. Many of these can be replicated in diverse contexts and may come to enjoy a virtually universal significance. These form the building blocks of place- and path-dependent forms of development. These include the joint stock, limited liability firm, the multinational corporation, the business union, the closed shop, and new firm spatial divisions of labor. They may originate in particular places and times and diffuse out to enjoy a more universal currency. At the present time there is widespread interest in the advanced capitalist countries in the adoption of Japanese production techniques simply and obviously due to the competitive edge they seem to have allowed. Other instances include the diffusion throughout the mining industry and much commercial agriculture in southern Africa of a particular set of labor control mechanisms including the pass, migrant labor and the compound.

Yet this fact of diffusion allows us to put a slightly different shine on the question of capital's geographical and historical differentiation. Capital's production logic and its globalizing tendencies can be read too much as constraint and insufficiently as opportunity. In particular too much focus on its disciplining tendencies can marginalize the question of the formation of new place-specific, path-dependent, technological-organizational ensembles. Storper and Walker (1989) have provided one angle on this question with their ideas of leapfrogging, windows of locational opportunity, and development *ab initio*. Their argument is that new technologies, embracing not merely new process technologies but new products and forms of organization, may be the basis of super-profits of a magnitude that allows a high level of locational discretion and, in particular, an independence of existing centers of capitalist industrialization. This may be because the new-ness of what is being done demands new skills, new sorts of components that are not available elsewhere; and/or because the super-profits being appropriated make the center a pole for skilled workers, component firms and the like so that it can create through its market power, its own territorial production complex. Lipietz has dramatized what is involved by referring to how "Storper and Walker, with almost Nietzschean accents, propose(d) an emerging model of 'poles of growth' arising from almost nothing" (1993, p. 13).

The point to be made here, however, is different and builds on Doreen Massey's (1983, 1992) work on place differentiation. Her starting point is to underline what she calls :

> [the] element of "chaos" which is intrinsic to the spatial. For although the location of each (or a set) of a number of phenomena may be directly caused (we know why X is here and Y is there), the spatial positioning of one in relation to the other (X's location in relation to Y) may not be directly caused. Such relative locations are produced out of the independent operation of separate determinations. . . . Thus, the chaos of the spatial results from the happenstance juxtapositions, the accidental separations, the often paradoxical nature of the spatial arrangements that result from the operation of all these causalities (Massey, 1992, p. 81).

But, she then goes on to argue, this chaos, these accidental juxtapositions, can have emergent powers; or in other words, provide the preconditions for new organizations and structures of social relations with their own distinctive causal properties.

It is not difficult to see the possibilities here. The co-location of firms with similar sorts of specialization is often quite accidental. But that co-location in turn allows the achievement of thresholds for a variety of external economies. The resultant powers then become necessary to the continuing competitive success of the firms and so lock them into that particular place where the necessary conditions for them are to be found. This place-embeddedness may be furthered to the extent that similar firms elsewhere and the locational advantages they provide are squeezed out or forced to relocate into this emergent complex of activity. The co-location of complementary activities like a consumer goods sector and the machine tool industry serving it might also become necessary from the standpoint of both, to the extent that is, that they develop relationships through which they can collaborate on the design and testing of new machinery and so become mutually dependent.

In other instances it may be accidental juxtapositions of the sort envisaged by Massey that allow firms to take advantage of more standardized types of innovative technological ensemble. An illustration of this is provided by a recent study of the engineering and metal working industries in the Lyons area of France (Lorenz, 1993). The focus there is the creation of a flexible production system in which engineering firms subcontract for certain parts to metal-working firms. Historically, the relation between the two types of firm had been more distanced: the engineering firms had only relied on the metal firms for overflow production. What altered this was the advent of CNC machine tools which the engineering firms would not be able to amortize themselves. Subcontractors, on the other hand, could aggregate demands to make it worth their while. This set in motion a process of vertical disintegration and the creation of relations of mutual dependence between the engineering firms and the metal fabricators. In turn this led to the (intended) creation of relations of trust between the two and the development of codes of behavior to facilitate that trust. What had been a contingent relation resulting from a chance juxtaposition became a necessary relation and, one assumes, a relation which substantially enhanced the ability of the engineering firms to compete on wider markets and take advantage of new technologies.

It is also possible that the Storper-Walker concept of new place formations and differentiations through windows of locational opportunity needs to be supplemented by the sorts of contingent circumstances or fortuitous combinations to which Massey is referring. While from one angle the emergence of a new technological pole might appear independent of pre-existing, localized conditions, closer inspection may demonstrate the importance of taken-for-granted aspects of (e.g.) state regulatory policy, banking structure, or nation-specific forms of training and expertise. In this regard Lipietz's words acquire a new significance:

> There can only be a certain type of "regions which win" (or rather a certain fashion of winning for a region) within the framework of a certain type of national state . . . and these states will only "win" in international economic competition if they know how to create this type of "regions which win" (1993, p. 16).

An important implication of these arguments is the simply enormous variety of local arrangements for production, technical and social, that we can reasonably anticipate. There exist a very large spectrum, in other words, of concrete structures of social relations which lock agents into particular places and structure, in terms both of constraint and opportunity, their developmental trajectories. This is not to say that this differentiation is limitless. There may be geographical variations but they have to "work." To the extent that they do they may diffuse out and achieve a more global presence. Indeed, as Storper has argued:

> The development of the economy of capital, arguably the best-theorized part of Marx's work, originates in historically- and geographically-specific experiments, which are only subsequently totalized through adoption, diffusion, and enforcement in real capitalist societies. What I mean to say here is that the study of the locality and the particular are keys to understanding how the "general" structure of capital develops. The local makes the global at certain critical points in space and time. I would argue that capital is forever altered by the rise of Japan not simply revealed "in another instance" (1987, p. 420).

It should be added, though, that they are rarely appropriated in a pure form. Local circumstances affect the development of new forms of technology and associated social organization as Storper has pointed out. But the way in which those forms are adopted elsewhere as they achieve some competitive power depends, as will be shown in the next section, on local historical legacies and politics. What results from these, in accordance with Massey's logic of accidental juxtapositions and emergent powers, may be new models to be adopted and adapted elsewhere.

THE CONTEMPORARY CONJUNCTURE

So finally, it is useful to come to a more direct confrontation with the new technologies and associated forms of social organization that are insistently pressed as the vehicles of a new era of capitalist accumulation. Given the very broad backcloth that this author has devised in terms of how we might think critically about geographic and historical differentiations of capitalism, what is one to make of arguments about the exhaustion of the possibilities of Fordism and the advent of various alternatives including flexible accumulation and post-Fordism?

There can be no doubt that there has been change. Its precise form or forms and causalities are more debatable. Consider first how we should situate it. The Brenner-Glick critique of Regulation Theory is useful here but also has a significant defect. They argue that Regulation Theory privileges period-specific institutions or rules at the expense of the underlying and enduring capitalist structure of relations they are supposed to be regulating. Rather institutional change, like the laws of motion of capitalism which generate the search for appropriate organizational forms, is ongoing and spans periods. We might add here its cumulative character and the way in which it facilitates a progressive socialization and hence development of the forces of production. These would include major institutional developments like state sponsorship of formal education as well as changing corporate forms like the multinational corporation or the Taylorization highlighted by the Regulation Theorists in their discussion of Fordism. Moreover, the self-regulating character of capital is seriously underestimated by Regulation Theory and, conversely, the role of the state in regulation receives more emphasis than it strictly merits.

One of the problems of the Brenner and Glick argument, however, is that in their emphasis of the importance of what they call the social-property relations of capitalism they neglect its tendencies to crisis. Here one should argue that in understanding the contemporary period of restructuring we need to take into account the fact that we have been on the downside of one of the long waves of capitalism: a period of serious overaccumulation of capital subsequent to the long boom from 1945 to about 1970–75 (Castells, 1980; Clarke, 1988; Harvey, 1988; Mandel, 1978).

This squeezing of profits and concomitant heightening of competition has elicited a wide variety of attempts at restructuring. The major types are familiar enough. What we need to be more aware of, perhaps, are the various combinations that are emerging, and the way in which different restructuring strategies often co-exist within the same firm, sometimes as a risk-spreading strategy, and the substantial distance that commonly exists between the ideal types identified in the literature and actual firm practice. These strategies include:

a. Japanese lean production with its much-publicized emphasis on "just-in-time." This facilitates profitability by reducing capital in the form of buffer stocks; and capital in the form of waste that is reduced by quality control and the rapid communication of shortcomings upstream. Geographers have been interested in just-in-time production systems largely from the standpoint of their implications for the spatial structure of production. But the implications are considerably broader than this. To the extent that set-up times for machines are protracted then buffer stocks will be required. This results in the use of smaller machines with shorter setup times. Aside from reducing waste, quality control can also provide an edge in final markets. Partly for this reason emphasis is placed on a long learning curve for labor, and also minimizing recruiting error and labor turnover. For the same reason relations with subcontractors tend to be long term and characterized by ongoing consultation regarding manufacturing processes and product design. As Sayer (1989) has pointed out, the long-term nature of relations with subcontractors and workers also helps to keep new technologies within the firm and retard imitation by competitors.

b. Flexible specialization as conceived by Piore and Sabel and their followers is an ideal type worth recounting. A supposed fragmentation of final markets has led to an emphasis on the production of new, design-intensive, products and on the development of new forms of work organization and technology to cope with frequent product changes. The latter includes the adoption of new computer-controlled machinery which can adapt to different specifications of what it is producing and with minimal setup times. Adaptation to a regime of frequent product changes and to interaction with the new, more complex machinery also calls for a more polyvalent workforce with upgraded skills. As products change, moreover, so must inputs. While flexible machinery can take care of some of this, vertical disintegration of labor processes and a subsequent flexibility in purchases is also important.

c. Peripheralization of production refers to a strategy first popularized by those arguing for a New International Division of Labor. It is a model which relies for its necessary conditions upon the production of standardized goods, using de-skilled labor processes, organized along classically Fordist lines, and upon the development of the means of communication and transportation. The attraction of this model is, of course, the curtailment of labor costs which it facilitates. It clearly cannot be limited to Third World locations since pools of low cost labor often exist in more developed countries in backward regions, or among female homeworkers, or immigrants from the Third World. "Peripheralization," therefore, does not necessarily have any specifically geographical connotations. Indeed, it may be brought about by a process of re-negotiation of contracts in the Fordist heartlands.

These are emphatically ideal types. In their implementation they are commonly partial and hybrid. There are national biases among the more developed countries. Flexible specialization has been more easily embraced in the former FRG than in Britain (Lane, 1988). It has also been pointed out that the implementation of subcontracting in Britain has not been straightforward. There is an awareness of the advantages that the Japanese have gained in this way but an unwillingness to develop the necessary relations of trust, and opportunism is still rife (Thorburn and Takashima, 1993). Likewise, Germany is not Japan, and its competitive effectiveness has been based on different principles (Mair, 1991; Lane, 1988). Alternatively different (restructuring) strategies may be implemented by the same corporation simultaneously.

With respect to the latter point, Andrew Mair (1991) has provided some very useful materials in a paper entitled, provocatively, "The Japanization of VW?" Historically, the VW labor process has been highly Taylorized; productivity improvements have been achieved by investing in hard technology managed by the skilled production workers supplied by Germany's renowned apprenticeship system. This recipe is now being challenged by Japanese production methods. This is dramatized by VW's withdrawal from production in the U.S. at the same time as the Japanese were (successfully) establishing transplants there. Japanization is on the agenda at VW and Mair has tried to document some of the initiatives undertaken under that rubric in a series of case studies (Mair, 1991).

Success hitherto is highly qualified. Attempts to implement just-in-time are seriously compromised by the off-center location of VW's main plants relative to those of components producers in Germany. The attempt to implement group work and so move away from Taylorization in the direction of a more polyvalent workforce has encountered difficulties. In this context the shift of VW into eastern Europe begins to look more and more like an insurance policy: the prolongation of its traditional Fordist labor process through the use of cheap labor. As Mair argues:

> One key finding from the case studies is simply the complexity of the situation "on the ground." This point is worth stressing because of a current tendency to oversimplify the process of industrial restructuring in the face of the Japanese challenge. . . . (p. 78).

The empirical variability and complexity of restructuring initiatives is also brought out by the literature on industrial districts. Flexible accumulation points in a clear direction as to the form these can be expected to assume, as in Allen Scott's (1988) concept of "new industrial spaces." But the "resurgence of regional economies" (Storper, 1994) appears to be more complex than this. Gertler has highlighted the issue:

> . . . how do we recognize a true "industrial district" or "territorial production complex" when we see one? How big or small, in terms of areal extent, can these be? What proportion of transactions involving producers within such districts need be wholly contained within the same area? And what kinds of social relations between producers in the area in question qualify as being representative of the unique behavior believed to take place within such districts? (1992, p. 263).

One of the instances he uses to make the point, and after Saxenian, is the contrast between Silicon Valley and Route 128 in the Boston area. In Silicon Valley the computer industry is more vertically disintegrated, trust-based and information-sharing and this has, perhaps, enabled it to respond more successfully to Japanese competition. Some sense of the variability existing across what are defensibly "industrial districts" is also apparent in Storper and Harrison (1991). Larger firms are far from absent. Smaller firms may cluster around them in tiers of subcontractors and sub-subcontractors. Alternatively, as Amin and Thrift (1992) have recently outlined, industrial districts at the cutting edge of a technology attract external investments by major multinational corporations struggling to keep abreast of emerging global standards. Industrial districts may also be the site of vertical integration prior to standardization and relocation elsewhere.

Some of the variety possible in the spatial structures associated with new production technologies are underlined by Andrew Mair's work on the implementation of just-in-time supply systems. Reference has already been made to imperfections in implementation by VW. In another paper (1992a) Mair has examined contrasting degrees of implementation in the heartland of "lean production." This varies from substantial in the case of Toyota, with its tiers of subcontractors and sub-subcontractors focused around Toyota City; to weak, in the case of Nissan.

Understanding these particular contrasts serves as an entrée into the more general question of coming to terms with this empirical variety of restructuring strategies *tout court.* In short, history, or, more accurately, historical geography as in Massey's chance juxtapositions, matters. For a variety of historical reasons all Toyota assembly plants are located together (in Toyota City) and surrounded by phalanxes of subcontractors, many of whom were encouraged to locate in Toyota City or nearby. The fact that Toyota dominates politics in Toyota City has been helpful in providing the physical infrastructure to facilitate the implementation of just-in-time. Less obvious are some labor factors. At the time of its establishment, Toyota City was in a relatively isolated, rural area and provided workers quite amenable to the demands that just-in-time makes on line workers: assignment flexibility and short notice overtime work in particular. To this day, Toyota has persisted in its practice of recruiting from rural areas.

Nissan is different. Originally located in the heart of Tokyo, new plants have been established on the edge of the metropolitan area. Many components suppliers have remained in the core. Part of its problem in implementing just-in-time is the extreme traffic congestion of the area. Another is the more militant character of the workforce which may have something to do with its dominantly urban character. At any rate, it has made it more difficult to persuade workers to do the overtime work commonly associated with just-in-time systems. This means that Nissan has to maintain higher levels of buffer stocks.

Context, it would seem, matters as to whether or not and/or the way in which new production technologies (again, in the broadest sense) can be effected. As Mair underlines, inherited geographies are not simply restructured. Market power can structure the location decisions of supplier firms, as has been to some degree the case of Honda in the U.S., but most firms do not enjoy that sort of power. On the other hand, unique configurations can also provide opportunities. It is no secret that successive British governments over the last 10 years or so have pinned a good deal of hope for restructuring British industry on Japanese investment. The idea is that direct investment by Japanese corporations in consumer durables would act as a vehicle for the introduction of new production technologies and concomitant organizational features. That this has met with some success is owing to circumstances of a quite contingent character. These certainly include Britain's relatively low wage costs and the loss of union strength that has accompanied de-industrialization. But of equal if not more significance is its position within the European Economic Community coupled with a decimated domestic auto industry that was in no position to mount the sorts of protectionism emanating from France and Italy: barriers, incidentally, that are sure to be dismantled as the EEC moves toward a removal of all obstacles to competition between firms in different member countries. On the other hand, a location in Germany was inadvisable on the grounds of the serious competition the auto industry there could mount. On such considerations does the success of (localized) restructuring strategies depend.

However, broader, more institutional features are also, quite evidently, implicated and once more underline the importance of historical legacies. In this respect, Christel Lane (1988) has put together an excellent comparative study of the progress made in the adoption of flexible specialization in Britain and the former FRG. Apparently, progress has gone much further in the German case. This partly reflects the fact that Germany was always significantly less penetrated by Taylorist labor process strategies than Britain and the enduring importance there of a craft paradigm which has lent itself to multiskilling and technical understanding. The much vaunted apprenticeship training characteristic of the country includes the acquisition of knowledge and abilities in two or more areas as a result of systematic rotation. This has allowed flexible use of labor across boundaries. In Britain, on the other hand, and quite apart from the decay of the apprenticeship system there, the notion of craft has been somewhat different. The role of the labor unions in training and accreditation has meant that skill comes with a job territory and defence of that territory against the claims of other trades and the unskilled: an orientation which makes polyvalency difficult.

The integration of the new technology with broader skill profiles has also been enhanced in the German case by distinctive features of industrial relations. Unions are not organized along craft lines nor do they support demarcation between crafts. At the plant level it is the works council rather than the union that is the representative body and these are dominated by skilled workers who, fearing technological change less, are more likely to support it. At the same time German employment legislation has made it difficult for employers to solve the problem of fluctuating demand by casualizing the labor force. There is a high degree of *de facto* employment security and this, along with the strong involvement of workers councils in manpower regulation, compels employers to engage in more careful and long-term manpower planning and to deploy the workforce flexibly in both functional and numerical terms.

In Britain the contrast is quite stark. There is not the same reservoir of skilled workers on which to draw, and their skills are quite narrowly defined and circumscribed by union-enforced demarcation. In consequence, there have been serious difficulties in merging mechanical and electrical skills, something thought to be especially important for maintenance in the new technologies. Relative lack of employment security has made the unions more fearful of change while at the same time encouraging on the part of employers a casualization of labor policies and a substitution of numerical for functional flexibility (cf. Pollert, 1988). As Lane summarizes the situation:

> The example of Britain shows, contrary to the claims of Piore and Sabel, that countries without a pervasive craft ethos will not necessarily remain wedded to the old model of industrial organization. Instead managements in such countries are more likely to develop a hybrid strategy, combining a changed market orientation with a high-tech version of production along the old Taylorist lines. Although greater functional flexibility of labor and upskilling is practised, the pursuit of numerical flexibility and of the greater casualization of the labor force appears to be the more widely adopted alternative. Although numerical flexibility can successfully cope with uncertainties in demand of a quantitative type it is doubtful whether a casualized labor force can handle other aspects of market uncertainty, such as frequent product changes or product diversification, and the demand for customized, high-quality goods (pp. 164–5).

But if history matters, and the legacy of national institutions in particular, then so too does politics. Institutional regimes are always points of balance between opposing political forces. Simon Clarke has argued that the rising mass consumption, growing social provision and inflationary expansion of credit of the post-war period should be seen not so much as aspects of a mode of regulation but as "related aspects of the institutionalization of a particular balance of class forces in the post-war period in which rising wages and rising levels of social expenditure were the price capital paid for the industrial and political integration of the working class in the immediate post-war period" (1988, p. 85).

More closely argued is Kenney and Florida's (1988) discussion of the political origins of the institutional context which facilitated the development in Japan of what they describe as post-Fordism. Guarantees of long-term employment and enterprise unionism did not just happen in some contractarian sort of way. Rather postwar class struggles are important in understanding what happened. Factory occupations and attempts to produce and sell the output led to the goal of establishing factory soviets. Enterprise unions uniting all workers emerged as the vehicle for accomplishing this. The granting of employment security was embraced as an important quid pro quo in abandoning more radical ends. The logic of the enterprise union underscored this demand since without employment security union membership and finances would have been in serious jeopardy.

The institutional regimes thus arrived at may accomodate new technologies with substantial facility, as in the German case reviewed above. In others they will be contested or, in a context of circumspection and fear for the political consequences, adopted in a tentative, piecemeal fashion. The British case seems to fit this one. This is not to say that a point of balance has been achieved. There are ongoing struggles to, for example, change the labor relations climate so as to allow for enhanced levels of functional flexibility. In an environment of serious unemployment Nissan was able to achieve single-union representation at its plant in Sunderland as a precondition, from its standpoint, of introducing more flexibility in worker assignment within the plant (Holloway, 1987). In a discussion of the strategic predicaments of Ford Motor Company in Europe, Mair has commented on its use of uncertainty as to the future of its operations in Britain in an effort to gain "the employee cooperation for changes to manufacturing the Japan-inspired model shows to be necessary . . . " (1992b, p. 160).

The politics of restructuring, however, is more complex than this. It is not just politics within particular arenas. Restructuring is being engaged in for competition on a global marketplace and within the context of capitalist relations of production. One needs to be particularly skeptical of the claims of Piore and Sabel in this regard. The structural logic of capital is such that the developments in the labor process that they point to cannot possibly become generalized.

The embrace of flexible specialization represents a strategy of strong competition: an attempt to achieve super-profits through the exploitation of new markets, the development of new products, and economies in the labor process. Like other strategies of strong competition it tends to be locationally embedded and so significantly place-dependent. Empirically this is obvious, as in the cases of Silicon Valley and the Third Italy. This is for diverse reasons including the flexibility achieved through vertical disintegration and the skills necessitated by flexible forms of manufacturing. As a form of strong competition, however, it only makes sense against the background of weak forms: the alternative, that is, of having profitability eroded to paper thin margins by the competition of other firms.

Weak forms exist as a result of the attempt to cheapen production so as to widen markets. There is no evidence to suggest that the products emanating from flexible specialization will not encounter the same fate though the heartlands of flexible specialization may avoid the consequences of it if they can develop a continuing stream of new products. In fact it will be a necessary development unless flexible specialization and mass production are to congeal into stable forms with their own market segments: an unlikely outcome given the underlying context of inter-firm competition. Consumer goods markets are stratified and necessarily so. Design-intensive, high-quality, radically new, products are more expensive to produce and their market penetration is necessarily limited. Given the downward pressure on wages that capital continually and necessarily exerts, demand for the standardized product using standardized inputs cannot but remain strong. This is the history of capitalist markets: there have always been high-quality, design-intensive goods around but most shopping baskets have been filled with standard, mass-produced goods like soda pop, chicken, beer, bread, detergents, work shirts, frozen foods, and so on. There may be microbreweries but their production pales beside that of the major brewers just as the purchase of Audis or BMWs has to be set by the side of that of Geo Prizms and Ford Fiestas. Strong forms of competition with their place- and path-dependent forms of development have, as their necessary complement therefore, the peripheralization of production.

The resultant politics of restructuring, therefore, is and cannot possibly be, anything other than global in scope. Firms and localities may want, desperately want, to become high technology-based; but for firms it is likely to be mass production that is in their future just as for localities the future holds branch plants and back offices. And as the stakes are heightened in this way so too is the political temperature.

CONCLUDING COMMENTS

The debate about flexible specialization raises important questions about period and place in our understandiing of capitalist development. Periodization is an issue either in the Piore-Sabel terms of successions of technological paradigms; or in terms of Regulation Theory and the question of what comes after Fordism. "Place" has entered into the picture, as rightly it should, at different scales. At the small scale there is the debate that has raged over the existence of new industrial districts or new industrial spaces as some have called them. And in the search for flexible specialization the nation state has acquired an explanatory status.

In assessing the explanatory claims of different periodizations and place-partitions, however, we need to hold to a clear and argued-out conception of the logic of capitalism. One is, after all, talking about periods and places in which that logic holds sway. The logic of capitalism may, as indeed has been argued here, substantially blur the neat discontinuities of time and space as well as cast doubt on the causalities that have been attributed to them. On the other hand, this author believes that in contrast to the infinite substitutabilities between locations in time and space implicit in the necessities of the capitalist mode of production there are serious non-substitutabilities, and has tried here to capture these with the (related) ideas of place- and path-dependence.

For particular technological-social structural ensembles do come into existence at particular points in time and space and lock agents into places and into particular paths of institutional and technical development. The precise content of these ensembles is difficult to predict. Their durability depends on their ability to bestow a competitive edge in an increasingly global market. They may diffuse and acquire a more universal currency as in the spread of lean production out from Japan. But they will also be altered in their encounter with different environments, in Massey's (1983) terms the deposits of past layers of investment, elsewhere. Indeed, this encounter may be the condition for new structures of social relations which provide those in particular places with new powers in the way Massey envisages with her idea of unplanned juxtapositions and the possibility of emergent powers.

It is, therefore, unreasonable to expect serious convergence between places in the late twentieth century in terms of the institutional-technical ensembles dominant there. Alongside flexible specialization there are other restructuring strategies like Japanese lean production which is flexible in some ways but quite inflexible in others. The degree to which these can be implemented depends on historical legacies; 'just-in-time' as a relation between firms and their suppliers is more feasible under some conditions than others. It also depends on politics. New forms of production are not necessarily for everybody and given the capitalist social structure which is their necessary context it would be surprising if they were. This is a politics, morever, that will be global. For to the extent that they endow competitive advantage for capitalist firms, the stakes as to their adoption will be enhanced.

REFERENCES

Amin, A, and Thrift, N. (1992). "Neo-Marshallian Nodes in Global Networks," *International Journal of Urban and Regional Research* 16: pp. 571–87.

Brenner, R. and Glick, M. (1991). "The Regulation Approach: Theory and History," *New Left Review* 188: pp. 45–120.

Castells, M. (1980). *The Economic Crisis and American Society*. Princeton: Princeton University Press.

Clarke, S. (1988). "Overaccumulation Class Struggle and the Regulation Approach," *Capital and Class* 36: pp. 59–92.

Cox, K. R. and Mair, A.J. (1989). "Levels of Abstraction in Locality Studies," *Antipode* 21: pp. 12–32.

_____. (1989b). "Urban Growth Machines and the Politics of Local Economic Development," *International Journal of Urban and Regional Research* 13: pp. 137–46.

Curry, J. (1993). "The Flexibility Fetish: A Review Essay on Flexible Specialization," *Capital and Class* 50: pp. 99–126.

Gertler, M.S. (1992). "Flexibility Revisited: Districts, Nation-States, and the Forces of Production," *Transactions of the Institute of British Geographers NS* 17: pp. 259–78.

Holloway, J. (1987). "The Red Rose of Nissan," *Capital and Class* 32: pp. 142–64.

Harvey, D. (1982). *The Limits to Capital*. Chicago: University of Chicago Press.

_____ . (1988). "The Geographical and Geopolitical Consequences of theTransition from Fordist to Fexible Accumulation." In G. Sternlieb and J. W.Hughes (Eds.), *America's New Market Geography*. New Brunswick: Rutgers University (Center for Urban Policy Research).

Kenney, M. and Florida. R. (1988). "Beyond Mass Production: Production and the Labor Process in Japan," *Politics and Society* 16: pp. 121–58.

Lane, C. (1988) "Industrial Change in Europe: The Pursuit of Flexible Specialisation in Britain and West Germany," *Work, Employment and Society* 2: pp. 141–68.

Lipietz, A. (1987). *Mirages and Miracles.* London: Verso Press.

_____. (1993). "The Local and the Global: Regional Individuality or Interregionalism," *Transactions of the Institute of British Geographers* 18: pp. 8–18.

Lorenz, E. (1993). "Flexible Production Systems and the Social Construction of Trust," *Politics and Society* 21: pp. 207–324.

Mair, A. (1991). "The Japanization of Volkswagen?" Local Government Centre, Warwick Business School, University of Warwick, Coventry, England.

_____ . (1992a). "Just-in-time Manufacturing and the Spatial Structure of the Automobile Industry: Lessons from Japan," *Tijdschrift voor Econmische en Sociaale Geografie* 83: pp. 80–92.

_____ . (1992b). "Ford of Britain in the 1990s: Corporate Strategy in Europe and the Future of the Ford Localities," *Local Economy* 7: pp. 146–62.

Mandel, E. (1978). *Late Capitalism.* London: Verso.

Markusen, A. (1991). "The Military-Industrial Divide," *Environment and Planning D: Society and Space* 9: pp. 391–416.

Massey. D. (1983). "Industrial Restructuring as Class Restructuring: Production Decentralization and Local Uniqueness," *Regional Studies* 17: pp. 73–89.

_____ . (1992). "Politics and Space/time," *New Left Review* 196: pp. 65–84.

North, D.C. (1990). *Institutions, Institutional Change and Economic Performance.* Cambridge: Cambridge University Press.

Piore, M J. and Sabel, C.F. (1984). *The Second Industrial Divide.* New York: Basic Books.

Pollert, A. (1988). "Dismantling Flexibility," *Capital and Class* 34: pp. 42–75.

Sayer, R.A. (1989). "Postfordism in Question," *International Journal of Urban and Regional Research* 13: pp. 666–95.

_____. (1992). *Method in Social Science.* London: Routledge.

Scott, A.J. (1988). "Flexible Production Systems and Regional Development: The Rise of New Industrial Spaces in North America and Western Europe," *International Journal of Urban and Regional Research* 12: pp. 171–85.

Storper, M. (1987). "The Post-Enlightenment Challenge to Marxist Urban Studies," *Environment and Planning D: Society and Space* 5: pp. 418–426.

_____ . (1994) "The Resurgence of Regional Economies, Ten Years Later: The Region as a Nexus of Untraded Interdependencies." (Paper presented at a Conference on Cities, Enterprises and Society at the Eve of the Twenty-first Century, Lille).

Storper, M. and Harrison, B. (1991). "Flexibility, Hierarchy and Regional Development: The Changing Structure of Industrial Production Systems and their Forms of Governance in the 1990s," *Research Policy* 21: pp. 1–16.

Storper, M. and Walker, R.A. (1989). *The Capitalist Imperative: Territory, Technology and Industrial Growth.* Oxford: Basil Blackwell.

Thoburn, J.T. and Takashima, M. (1993). "Improving British Industrial Performance: Lessons from Japanese Subcontracting," *National Westminster Bank Quarterly Review.* February 2–12.

Walker, R.A. and Large, D.B. (1975). "The Economics of Energy Extravagance," *Ecology Law Quarterly* 4: pp. 962–85.

Weeks, J. (1981). *Capital and Exploitation.* London: Edward Arnold.

Williams, K., et al. (1987). "The End of Mass Production?", *Economy and Society* 16: pp. 405–39.

11

THE CURRENT TRANSITION RECONSIDERED

Daniel C. Knudsen
Department of Geography and Institute for Development Strategies, School of Public and Environmental Affairs, Indiana University, Bloomington, IN 47405 USA
and
Raymond L. Martin
Department of Geography, Indiana University, Bloomington, IN 47405 USA

INTRODUCTION

This volume has presented collected essays on various aspects of the current transformation of industrial capitalism in an attempt to create a single complete picture. The book touched upon five dimensions of the current transformation of capitalism: technology, labor relations, firm organization, public policy, and locational decisionmaking. The purpose of this final chapter is to summarize the debate contained within the book and to extend the findings beyond the narrow confines of the largely Anglo-American and U.K. experience that the authors present.

SUMMARY OF THE ESSAYS CONTAINED IN THE BOOK

Summarizing the essays is most directly accomplished by paraphrasing the conclusions drawn by the authors themselves. In the introduction, *Knudsen and Boggs* argue that the advanced capitalist nations are currently undergoing a wrenching economic, social, and political transformation. At its heart, this transformation involves a shift away from the mass production of highly standardized goods and services for mass markets to small-batch production of relatively customized goods and services for niche markets. Technologically, the introduction of new "flexible manufacturing" techniques have increased the ability of firms to respond to the very particular and frequently changing demands of niche markets. Flexible manufacturing is most often associated with the use of sophisticated computer controlled machine tools that simultaneously allow high-speed production, rapid changeover between product lines, and high quality. New information technologies have allowed the integration and coordination of activities both within and across firms and this increased integration and coordination has led to phenomenon such as "just-in-time" production.

The introduction of flexible machine tool technology has often gone hand-in-hand with alterations in the "society" of work. The increased horizontal authority of existing managers and worker empowerment are two of the more visible aspects of this alteration. These changes in the "society" of work have led to dramatic reductions in both the number of production workers (the recession of 1979–83) and the number of middle-level managers (the recession of 1990–91), and have substantially increased the contingency of the labor force. The transition to flexibility also has altered the area of interfirm relations, spawning a surprising variety of new interfirm relationships. These forms span the spectrum from that anticipated by Piore and Sabel (Italianate and neo-Marshallian) to those relatively similar to those that prevailed under Fordism. Thus, it is not uncommon to see small, vertically disintegrated but highly networked firms operating in the shadow of large, vertically integrated producers.

Malecki notes that in the current context flexibility, technology, and competitiveness are highly interrelated. Firms with a larger number of sources of information about markets, competitors, and new techniques introduce a greater number of new products or services. These sources may be mainly within the local base of firms, or they could be based on wide-ranging external networks. Firms with external networks are more apt to be innovative. Some suggest that the flexible firm has been superseded by the innovative or the agile firm. Such firms are multi-dimensional and multi-talented, combining innovativeness, efficiency, quality, and flexibility, while retaining strengths in its core competencies.

Kenney and Curry highlight a major contradiction in emerging labor relations in the contemporary global economy. This contradiction is the simultaneous co-existence of two apparently divergent poles of labor organization. It is possible to identify three reasons for this bifurcation in labor organization and for the persistence of low skill, low wage labor in particular. First, some processes are extremely hard to automate (for instance, garment and some assembly operations) because the physical form of the product and the methods of manufacture are not amenable to machine production. Second, for some processes automation is undesirable (restaurants, entertainment, etc.). The third instance consists of processes which are inherently knowledge-intensive and cannot be automated or routinized, or done so only in a limited fashion (e.g., primarily creative activities such as software writing, fashion design, marketing, management and coordination, production of entertainment goods such as books, movies, music, etc.).

Mikulecky argues that basic skill demands in the workplace have both expanded and increased. Today's workplace requires more flexibility, cooperation, and independent competence of employees. Recent surveys indicate that, while, on average, workers appear to have the capabilities to perform complex tasks called for in the new workplace, nearly every occupation has a significant portion of current employees who cannot consistently and accurately perform these complex tasks. In several occupational classifications, more than half of all workers are incapable of consistently performing tasks associated with restructured workplaces. Similarly, schools appear to have made little headway in educating students for the modern workplace. While fewer students have less than minimal skills, fewer students excel. Minority students and those from low-income homes are over-represented among students with low skill levels. There are indications that the gap between school and workplace is now beginning to be recognized and that schools and employers are responding to this gap. New programs, such as the Tech Prep program appear promising, but have so far affected few students and adults.

According to *Pavlik*, the current patterns of production by large firms represent both change and continuity. Unlike patterns of production that emerged in the immediate postwar period, when large, vertically integrated firms followed strategies that eventually would be incorporated into the product cycle and new international division of labor models, current strategies are more tied to the nature of the industry in question and the degree of customization of the product. This has led to a bifurcation in the strategies that firms employ, and even in differing strategies within separate branches of the same firm. The result has been both a reconcentration of production in certain regions, and the development of spatially extensive supply and production networks.

Warf remarks that services play an important role in the growing flexibility of capitalist enterprises to deal with much higher levels of uncertainty and rapidity of change than in the past. The service economy emerged long before the birth of flexibility. The reasons for the recent rapid growth of services lay in the changing incomes, demographic structure, and levels of public intervention in western societies. However, services would not be exchanged to the degree that they are today without benefit of flexible production systems. Manufacturing and services, each of which embraces a variety of production techniques, are contingent but mutually reinforcing systems, reflecting and reinforcing the evolution of commodity production. Services also have increasingly incorporated numerous flexible production technologies. These new technologies have radically realigned the nature of work within producer service industries and their geographical location. The introduction of electronic funds transfer systems, for example, have given banks, securities, and insurance firms great freedom over their locational choices, enabling a separation of front and back offices and tremendous growth in the offshore banking industry. By dramatically reducing the circulation time of capital, telecommunications have not eliminated differences between places, but rather have permitted the exploitation of those differences. Lastly, the digitization of information and the introduction of telecommunications has resulted in services no longer being easily dismissed as a parasitic, locally consumed activity. Mounting evidence indicates that they are traded interregionally and internationally.

Bosworth argues that successful firms are those that network and form strategic alliances. Within the current global competitive situation, such strategies are crucial to survival, particularly for small firms. Given that the principal impediments to survival for firms are those associated with the acquisition of knowledge about new markets, new processes, and new products, learning is extremely important to competitiveness. Few firms, and especially few small firms, are capable of amassing the necessary resources to effectively learn. Networks and strategic alliances are a crucial tool in the hands of firms to both more effectively receive and use information, and to learn. From an economic development stance, helping firms effectively use information and learn is crucial because firms with these attributes are often globally competitive. Globally competitive firms are "high value-added firms," and high value-added firms are important for boosting the standard of living. Too often public policy has focussed on individual firms, not interconnected networks of firms. Public policy directed at individual firms is both inefficient and, typically, ineffective.

Wheeler notes that government can be instrumental in laying the foundation for a more adaptive and collaborative economy by supporting moves toward flexibility and change while protecting important social and political goals. At the same time, it is critical that government also re-engineer itself to be more flexible, and to utilize more performance and market-based metrics of success for service delivery. Finally, while it is the legitimate role of governments to provide transitional adjustment assistance, protecting firms from having to change is both socially and financially costly and doomed to failure.

Leigh focusses on an array of trends that are affecting the use of the built environment (fixed structures, specifically, office and industrial facilities) in what is perceived as a heightened era of flexibility in the capitalist economy. Current trends do not bode well for the fate of central city economies. There is little that suggests that the tide of abandonment of older office and industrial buildings will be turned around. Industrial capitalism has long relied on its ability to freely shift locations, abandoning redundant sites for new sites, consuming increasing amounts of greenfield land in the process. In the late twentieth century, we are increasingly forced to contend with this downside of capitalist production. The negative externalities associated with abandoned buildings include increased vandalism, physical deterioration, lower property tax income, and (in many cases) environmental contamination. The lack of economic use of such buildings decreases employment opportunities for the area's population, particularly if that population is low-income and has limited mobility. The resulting extended pattern of suburbanization increases pressure on already scarce public resources to fund new suburban infrastructure improvements and decreases that available for urban infrastructure redevelopment. In our present regulatory and real estate financial climate, there is little or no incentive for reusing existing infrastructure.

According to *Cox*, the debate about flexible specialization or flexible accumulation raises important questions about period and place in our understanding of capitalist development. In assessing the explanatory claims of different periodizations and place-partitions, however, we need to hold to a clear and argued-out conception of the logic of capitalism. The logic of capitalism may substantially blur the neat discontinuities of time and space as well as cast doubt on the causalities that have been attributed to them.

While infinite substitutabilities between locations in time and space exist, serious non-substitutabilities exist as well. These may be captured using notions of place- and path-dependence. Particular technological-social structural ensembles do come into existence at particular points in time and space and do lock agents into places and particular paths of institutional and technical development. The precise content of these ensembles is difficult to predict. Their durability depends on their ability to bestow a competitive edge in an increasingly global market. They may diffuse and acquire a more universal currency as in the spread of lean production out from Japan. But they will also be altered in their encounter with different environments. Indeed, this encounter may be the condition for new structures of social relations which provide those in particular places with new powers in the way Massey envisages with her idea of unplanned juxtapositions and the possibility of emergent powers. It is, therefore, unreasonable to expect serious convergence between places in the late twentieth century in terms of the institutional-technical ensembles dominant there. Alongside flexible specialization there are other restructuring strategies like Japanese lean production which is flexible in some ways but quite inflexible in others. The degree to which these can be implemented depends on historical legacies and on politics. New forms of production are not necessarily for everybody and given the capitalist social structure which is their necessary context it would be surprising if they were. This is a politics, moreover, that will be global.

CONCLUSIONS

It is tempting to highlight the differences between the U.S. and U.K. experiences and those of the remaining advanced capitalist nations. Clearly, there is some precedence for such a distinction. For example, the literature on international political economy regularly distinguishes between the "neo-liberal," "corporatist," and "negotiative" approaches to the recent transition (Amin and Thrift, 1995; Jessop, Nielsen, and Pedersen, 1993). However, as Cox points out in the previous chapter, these sort of artificially sharp distinctions may be no more informative than the artificial distinction between Fordism and flexibilism. There exists, for example, almost as much difference between the two countries most typically identified with the corporatist approach, Germany and Japan, as there is between the neo-liberal and corporatist approaches. German labor relations revolve around state-brokered negotiations between the industrialist's federation BDA (Federation of German Employers' Association) and the DGB (Federation of German Trade Unions) that represents 7.5 million German workers across 17 separate unions. Thus, German labor is organized along craft union lines. In Japan, labor relations revolve around firm-brokered negotiations with (their own) enterprise unions. Firm organization is also considerably different. Germany is home to a particular variant of flexible production that has been labeled "flexible specialization" (Piore and Sabel, 1984). This form of organization represents a revival of craft-based production within a set of internally disintegrated firms. Japan, however, is home to a variant of flexible production that has been termed "lean production" and which arose as the unintended result of a state-supported effort to create a system of mass production

(Hart, 1992, p. 186). Lean production typically takes place within a system of highly tiered suppliers and subcontractors. Finally, the roles of the state are quite different in the two nations. German industrial policy, while aligned with organizations, such as the Bundesbank and the Ministry of Finance, has a distinctly regional character (see Hart, 1992). Japanese industrial policy is much more nationally focussed within the Ministry of International Trade and Industry, the Ministry of Finance, and the Bank of Japan (Hart, 1992, pp. 38–40).

Much of the discussion in this volume is couched in comparative terms. For example, the essay by Kenney and Curry compares Japan, the U.S., and Mexico, Bosworth draws primarily on European examples, and the essay by Cox discusses differences between the U.S., Germany, and the U.K. Further, the essays by Pavlik and Malecki are general and do not greatly depend on a particular national context. Thus many of the arguments contained within this volume, while framed within the neo-liberal experience, are a great deal more generalizable.

Nevertheless, one important difference appears between the neo-liberal and the corporatist nations. This regards the manifestation of the restructuring of labor relations. As Kenney and Curry argue, in the neo-liberal nations, the restructuring of labor relations has typically manifest itself as a bifurcation of the labor market into high-wage, high-skill employment on the one hand, and low-wage, low-skill employment on the other. This bifurcation is often (though not always) coincident with downward pressure on wages and overall job market growth. In most of the corporatist nations, social and political institutions have mitigated against downward pressure on wages. This has resulted in a less-pronounced bifurcation of the labor market and in a very slow rate of growth in the overall job market. The result in most corporatist nations, therefore, has been rising and/or high (structural) unemployment and an expansion of the informal, or "dark," labor market.

There continues to exist a divergence of opinion over whether the current period of industrialization marks an identifiable period of capitalism distinct from the past. This volume will no doubt add to that debate, but it has not been our task to resolve it. Still, it is clear that conditions have changed. While Fordist production is one answer to organizing production within a set of national economies (increasingly linked internationally) that can absorb virtually all that is produced, the move toward more flexible approaches to production is the response to a need by firms to create and organize demand in an age of overcapacity and global competition. Further, while the central attributes of current industrial capitalism, small batch production, labor contingency, and vertical disintegration are global tendencies, these global tendencies are articulated by national and regional histories and politics, so that there is no single "best" approach to flexibility.

REFERENCES

Amin, A. and Thrift, N. (1995). "Institutional Issues for the European Regions: From Markets and Plans to Socioeconomics and Powers of Association," *Economy and Society* (forthcoming).

Hart, J. (1992). *Rival Capitalists: International Competitiveness in the United States, Japan, and Western Europe.* Ithica, NY: Cornell University Press.

Jessop, B.; Nielsen, K; and Pedersen, O.K. (1993). "Structural Competitiveness and Strategic Capacities: Rethinking the State and International Capital." In S. Sjøstrand (Ed.), *Institutional Change: Theory and Empirical Findings*. New York: M.E. Sharpe.

Piore, M. and Sabel, C. (1984). *The Second Industrial Divide*. New York: Basic Books.

GPSR Compliance

The European Union's (EU) General Product Safety Regulation (GPSR) is a set of rules that requires consumer products to be safe and our obligations to ensure this.

If you have any concerns about our products, you can contact us on ProductSafety@springernature.com

In case Publisher is established outside the EU, the EU authorized representative is:

Springer Nature Customer Service Center GmbH
Europaplatz 3
69115 Heidelberg, Germany

Batch number: 10704043

Printed by Printforce, the Netherlands